THE RHYTHMIC MOVEMENT METHOD

*A Revolutionary Approach to Improved
Health and Well-Being*

HARALD BLOMBERG, MD

Copyright © 2015 Harald Blomberg, MD.

All rights reserved. No part of this book may be reproduced, stored, or transmitted by any means—whether auditory, graphic, mechanical, or electronic—without written permission of both publisher and author, except in the case of brief excerpts used in critical articles and reviews. Unauthorized reproduction of any part of this work is illegal and is punishable by law.

ISBN: 978-1-4834-2879-6 (sc)
ISBN: 978-1-4834-2878-9 (e)

Library of Congress Control Number: 2015904754

Because of the dynamic nature of the Internet, any web addresses or links contained in this book may have changed since publication and may no longer be valid. The views expressed in this work are solely those of the author and do not necessarily reflect the views of the publisher, and the publisher hereby disclaims any responsibility for them.

Any people depicted in stock imagery provided by Thinkstock are models, and such images are being used for illustrative purposes only. Certain stock imagery © Thinkstock.

For information about workshops in rhythmic movement training, visit www.blombergrmt.com.

For Your Information

The procedures and techniques described in this book are solely for educational purposes. This book is for instructional purposes only. The author does not, directly or indirectly, present any part of these works as a diagnosis or as a prescription for any difficulty or challenge for any reader or student. Persons using the activities, procedures, and movements reported herein do so for educational purposes only.

Lulu Publishing Services rev. date: 04/21/2015

CONTENTS

Preface vii
Introduction xi

Chapter 1   How Doctors Traditionally Treat ADHD 1
Chapter 2   An Alternative Way of Regarding and Treating ADHD 13
Chapter 3   Two Different Ways of Looking at Children with Challenges 26
Chapter 4   Environmental Causes of Attention and Learning Problems 42
Chapter 5   The Brain Stem and Rhythmic Movements 51
Chapter 6   The Cerebellum and Rhythmic Movements...................61
Chapter 7   The Reptilian Brain, or the Basal Ganglia......................69
Chapter 8   Primitive Reflexes Especially Important in ADHD......79
Chapter 9   Primitive Reflexes in Reading and Writing Difficulties 96
Chapter 10   The Limbic System and Rhythmic Movements 107
Chapter 11   The Prefrontal Cortex and Rhythmic Movements.....128
Chapter 12   Autism Spectrum Disorder and Rhythmic Movements 135
Chapter 13   Rhythmic Movement Training and Psychosis 151
Chapter 14   Food intolerance and rhythmic movement training ... 165
Chapter 15   What Is Dyslexia? 174
Chapter 16   Visual Challenges and Dyslexia 184
Chapter 17   Phonological and Writing Challenges 206
Chapter 18   The Neural Network of Reading and the Prefrontal Cortex 215

Appendix: Rhythmic Exercises 227
Glossary..247
CV of Harald Blomberg 253
References 259

PREFACE

This book is an updated version of my original Swedish book *Rörelser som helar* (2008), which was first published in English with Moira Dempsey under the title *Movements That Heal* (2011).

Rhythmic movement training (RMT) is not a subject that can be confined into a rigid frame. It needs to be continually expanded. Helping children to develop, mature, or heal physically, emotionally, and mentally has become much more challenging than it was when I first started to work with this method. In both my own experience and in the experiences of teachers and therapists I've met in my courses, children nowadays have many more problems than they used to have ten years ago—or even three or four years ago. There has been a dramatic increase in attention problems, autism, emotional problems, and learning difficulties.

More and more frequently, it happens that children with ADHD, learning difficulties, and motor problem do not make the swift progress with rhythmic movement training that I used to see when I first started to work with this method. Therefore, I have taken it as my task to investigate the causes of children's rapidly declining health and how it can be dealt with when working with RMT. I have enlarged the information about how the environment affects children's health and added a chapter about food intolerance and rhythmic movement training. Nowadays, the immune systems of all children—not only children with autism—are under severe stress due to radiation from microwaves, heavy metals, food additives, and other chemicals; and this has caused a dramatic increase of gluten sensitivity and celiac disease. Stress from mobile phones and cordless networks and from gluten and casein sensitivity is a significant cause of the rapidly declining state of health among children nowadays and needs to be dealt with in order to make RMT effective.

Consequently, rhythmic movement training should not only focus on movements and reflex integration but also needs to consider environmental factors in order to be truly helpful in the long run. Rhythmic movement training in combination with the steps I recommend in this book is truly a revolutionary approach to increased health and well-being. This is why I have taken the opportunity with this new edition to call the book *The Rhythmic Movement Method* with the subtitle *A Revolutionary Approach to Improved Health and Well-Being*.

This book is written for all people who want to find methods to help both adults and children with challenges in order to make them feel well, function well, and stop taking medications. The book describes simple healing exercises developed by Kerstin Linde that stimulate the ability of the brain and the nervous system to renew itself and create new nerve connections and how these exercises help the client to develop, mature, or heal physically, emotionally, and mentally. These movements exist as inborn motor patterns that need to be activated in order for children to develop normally. I have called the method rhythmic movement training (RMT).

The exercises may sometimes seem to have a magical effect within a short time; however, this does not mean that their manner of action is magic. I have done my best to explain in a scientific language how such simple exercises can have such powerful effects, and I hope that the reader will not feel weighed down by my explanations.

I learned the exercises from Kerstin Linde and have used them for more than twenty-five years in my work as a psychiatrist. Kerstin Linde originally worked as a photographer, and by observing how infants, children, and adults move, she worked out her method, which was primarily inspired by the rhythmic movement that infants spontaneously make. I followed her work for a few years in the end of the 1980s and have written about that in my previous book, *Helande Liv* (*Healing Life*), which has not been translated into English.

The rhythmic elements of movements are especially characteristic of her method. Other approaches to helping children with motor and learning challenges have also been inspired by baby movements but lack the rhythmic elements. As will be evident from this book, the rhythmic element of the spontaneous infant movements is of fundamental importance for the motor, emotional, and intellectual development of the child.

One of many effects of Kerstin Linde's method was the integration of infant or primitive reflexes. She used to emphasize that the rhythmic exercises integrate primitive reflexes if they are done in a correct or exact way. This may be true as to small children, but in older children and adults, the rhythmic exercises need to be supplemented with other methods. A very effective method to integrate primitive reflexes has been developed by Russian psychologist Svetlana Masgutova. Her exercises are a valuable supplement of the rhythmic exercises.

For many years, I have taught classes of rhythmic movement training and reflex integration in Sweden and other countries in Europe, also teaching in Asia, Australia, and America. In these courses, I have addressed myself to teachers, physiotherapists, occupational therapists, behavioral optometrists, masseurs, and other professionals as well as to parents of children with different challenges.

This book is based on the manuals of these courses and could be used to replace especially the theoretical parts of these manuals. It has been my purpose to give a plausible scientific explanation of the remarkable effectiveness of rhythmic movement training in so many different conditions. The book therefore contains somewhat elaborate sections about the structure and function of the brain and the nervous system and about primitive reflexes. It also contains many case reports that illustrate how the method works.

I want to thank Dr. Mårten Kalling for his assistance with many valuable insights and scientific articles that have helped me to explain the mode of action of rhythmic movement training.

For help with the illustrations in this book, I want to thank Sandra Almenberg and Ricardo Mauler Gruber.

Harald Blomberg
Stockholm
December 2014

INTRODUCTION

My meeting with Kerstin Linde

In 1985, I met Kerstin Linde, who had developed a method that she called rhythmic movement pedagogy. At that time, I worked as a psychiatrist in a psychiatric outpatient clinic and trained as a coach of neurolinguistic programming. I also took part in a two-year education course of hypnosis.

I was intrigued by Kerstin Linde's account of her work with children and adults using rhythmic exercises. She claimed to have excellent results working with all kinds of clients, from children with severe motor handicaps to adults with psychosis and depression. I asked her to allow me to sit in when she was working with clients in order to be able to learn and understand what she was doing, and she graciously allowed me to do so. She advised me to follow her work with handicapped children because that would be most instructive, so I did.

Nothing had prepared me for the progress I saw in many children with severe motor handicaps, least of all my medical training. Spastic children who could hardly move, were unable to talk, and had severe strabismus or farsightedness relaxed their muscles. After a few months, some of them were already able to crawl, rise along furniture, and even speak sentences with three or four words. I observed how strabismus could all but disappear and how the training could improve farsightedness considerably.

The parents of these children were as surprised as I was by the rapid progress of their children, especially since doctors and physiotherapists had told them that they should not expect any substantial improvement in their children from the treatment that was offered by the medical service. In many cases, parents felt that doctors and physiotherapists regarded it as more important to teach them to gain insight of their children's handicaps and accept status quo than help their children to get better.

Rhythmic movements with psychiatric outpatients

Encouraged by the improvements I observed—not only in children with severe handicaps but also in adults with problems such as back pain, osteoarthritis of the joints, or psychiatric symptoms—I introduced the rhythmic exercises in the outpatient clinic where I was a consultant. Patients learned a few simple rhythmic exercises that they were told to do once a day for no more than ten minutes. The exercises soon became quite popular among patients since many experienced relief of depression, anxiety, or psychotic symptoms.

I noticed that the exercises stimulated many patients to remember their dreams, and this was like opening a new world for some of them. I also noticed a psychological development in my patients, which in many cases was reflected in their dreams.

The nurses who attended to schizophrenic and psychotic patients noticed that they improved in many respects. They became less withdrawn, more active, and interested in socializing. The psychotic symptoms were reduced and even disappeared completely in two patients who had suffered from schizophrenia for several years.

The patients were grateful and happy for the treatment, but when my superior heard about it, he forbade me to continue this practice, arguing that the treatment "was not accepted or especially well known."

I refused to oblige, and in order to stop me, he had no other alternative but to report me to the National Board of Health. An investigation was started in 1988, and I wrote ten case studies documenting the effects of the treatment. Many of my patients wrote to the board, expressing their appreciation of the treatment. The board established in its final report that the treatment was "experienced very positively by many patients" and that the "movement treatment was a worthwhile contribution in a situation that had appeared to be deadlocked or stagnant."

Moreover, the board criticized my superiors for lack of coordination in the treatment of inpatients and outpatients. After this, I was totally rejected by my superiors, which made my work situation impossible, and I decided to resign.

A scientific study of the rhythmic exercises is initiated

In 1989, I started private practice. A year later, a colleague, Dr. Mårten Kalling, invited me to introduce the movement training for some severely ill chronic schizophrenic patients, most of who had been hospitalized for ten years or more at a psychiatric hospital. I started to work there two days a week. In 1991, I was offered a part in a research study of this work, supervised by an assistant professor of psychology at the University of Umeå.

I had to write an application to get a grant for the project, summarizing previous research in the same field and describing the mode of action of the rhythmic exercises and why they would be effective in schizophrenia. I found no research of any method that bore any resemblance to the rhythmic exercises, and I accordingly entered virgin soil when I tried to explain the mode of operation of the rhythmic exercises.

The research study was continued for two years and showed favorable results. The patients treated with the rhythmic movements had displayed the greatest positive change compared to a control group. They had become more interested in their surroundings, were able to take part in social activities, and were able to participate in occupational therapy and their daily tasks in the ward to a greater extent.

The mode of action of rhythmic movements according to Kerstin Linde

Kerstin Linde described the rhythmic exercises as rhythmic whole body movements. According to her, the method is based on a functional comprehensive view: By eliminating functional disorders of the body as a whole, the symptoms that have resulted will indirectly be corrected.

With the training, the brain learns to control the body and the motor organs and automatically apply the proper level of muscle tension necessary in each moment. The goal of the training, according to Kerstin Linde, is to ensure that circulation and exchange of gases—oxygen, CO_2, and so forth—function in all parts of the body.

She got her inspiration for the method from the rhythmic movements that infants spontaneously make before they rise and walk. Through these movements, infants learn to apply the proper muscle tension when they move and to manage gravity automatically. If we do not have a basic tuning of our muscles as infants, we may automatically apply muscle tensions that are detrimental for our joints and spine and/or obstruct circulation and exchange of gases. This may eventually cause pain and the wearing of our joints, especially the knees, hips, and spine.

A complementary explanation of the effects of the rhythmic exercises

Kerstin Linde's theories about the rhythmic movements were not very helpful in explaining how they can stimulate the development of language in children with CP or improve psychotic symptoms in chronic schizophrenia. I had to find other explanations for the efficiency of the rhythmic exercises in such cases.

I was inspired by the theory of the triune brain by Paul MacLean, according to which different levels of the brain are responsible for motor abilities, emotions, and cognitive functions. These parts of the brain have been set up in the newborn baby, but they are not yet fully developed and linked together. This should normally take place during the first years of life.

When I observed Kerstin Linde's work with severely motor handicapped children, I noticed that the more motor handicaps they had, the less they had developed other functions such as speech and emotional and cognitive functions. The more rapidly their motor abilities progressed, the more rapidly these functions also developed. This observation made me conclude that the brain needs stimulation from motor activity in order to develop and mature and that such stimulation links the different levels of the brain together. However, brain researchers and medical doctors in general do not seem to recognize this fact, believing that the brain only needs oxygen and nutrition and develops like a head of cabbage.

I was also able to formulate a plausible explanation as to why the rhythmic exercises improve both language and psychotic symptoms. I later explained this theory extensively in my first book about rhythmic movement training, which was published in Swedish in 1998.

Rhythmic movements and primitive reflexes

Before I met Kerstin Linde, I had already attended a course of primitive reflexes and learning disability taught by Peter Blythe, founder of the Institute of Neuro-Physiological Psychology (INPP).

Primitive reflexes are automatic stereotyped movements controlled from the brain stem. These reflexes manage motor activity of the fetus and the newborn infant, and they must be inhibited and integrated for the motor ability of the child to develop properly. The infant integrates the primitive reflexes by making rhythmic infant movements that repeat the pattern of the different reflexes. Kerstin Linde used to say that she could observe primitive reflexes but did not have to work with them specifically since they were integrated with the rhythmic exercises she used.

In 1994, I started to work full time in my private practice. Especially when I used the rhythmic exercises I learned from Kerstin Linde in my work with children, I observed that all of them were effective for integrating primitive reflexes. I also found that some of the exercises similar to infant movements actually could integrate several primitive reflexes.

Around ten years ago (at the beginning of 2000), I learned another way of integrating primitive reflexes by attending courses taught by Russian psychologist Svetlana Masgutova. Her method was to reinforce the reflex pattern with a slight isometric pressure, a method that was especially useful in older children and adults.

Rhythmic movement training

During the nineties, I taught occasional courses of rhythmic movements for therapists, teachers, and nursing staff. After the publication of my first book in 1998, these courses increased in demand. In 2002, I started to teach courses regularly in Sweden.

I initially created three courses, each course corresponding to one level of the triune brain. In my first course, which mainly focused on the brain stem and the reptilian brain, I taught how and why the rhythmic exercises could be used not only to improve motor ability but also to improve attention and hyperactivity and to integrate primitive reflexes common in ADHD. In my second course, I focused on the mammalian brain, which corresponds to the limbic system and is responsible for emotions. I taught how the rhythmic exercises affect emotions and improve self-assurance and self-confidence. The third course focused on functions of the cortex and dealt with reading and writing difficulties and how to improve visual and phonological problems and reading comprehension with rhythmic exercises and special reflex integration exercises.

I did not want to limit myself to what I had learned by following Kerstin Linde's work and decided to include other topics in my courses, such as testing and integration of primitive reflexes. Besides using rhythmic exercises for integration of the reflexes, I also included exercises with isometric pressure. My extensive experience during more than fifteen years of using the rhythmic exercises with both children and adults suffering from a wide variety of complaints was an invaluable asset in creating these courses.

My aim was to explain the mode of actions of the rhythmic exercises plausibly in a scientific yet simple way that could be understood by ordinary readers without medical education. Dr. Mårten Kalling provided me with many scientific articles that helped me to do this.

In Sweden, only persons with medical training are allowed to treat children below the age of eight years. Kerstin Linde regarded the rhythmic movements as a teaching method and called her method pedagogy, not therapy. I decided to call the extended method I had created based on Kerstin Linde's rhythmic exercises rhythmic movement training, in Swedish *rytmisk rörelseträning*.

The further development of rhythmic movement training

I decided to offer to teach my courses to sponsors willing to engage me since I regarded myself primarily as creator and developer of rhythmic movement training and not as an organizer. In Sweden, a center for sensory integration and positive learning sponsored my courses.

In 2003, I started to cooperate with Moira Dempsey. I translated my Swedish manuals into English, and she edited and illustrated them as well as sponsored my courses in Southeast Asia, Australia, and the United States.

In 2005, I was invited to teach my courses in Spain, where rhythmic movement training has become widespread and popular. I regularly teach several courses a year there.

Besides teaching, I continued to work in my private practice, especially focusing on working with children with motor problems, attention and learning difficulties, and disorders within the autism spectrum. I continued developing my original courses, inspired by what I learned from working with children and by my experiences of teaching people with various backgrounds. In Spain, my courses became especially appreciated among optometrists, and recently a short course of rhythmic movement training taught by my Spanish sponsor, Eva Maria Rodriguez Diez, was included in their training at the University of Madrid. What I learned from Spanish optometrists helped me to improve my course about reading difficulties and especially to develop new and more effective exercises for visual problems.

New courses of rhythmic movement training

The rhythmic movements can successfully be used in many different areas, as Kerstin Linde already had shown. For the education of preschool teachers mentioned above, I created a special course teaching the rhythmic exercises with songs and nursery rhymes and using games to integrate the reflexes.

The rhythmic exercises are excellent for psychotherapy especially since they promote dreams and help people get in touch with subconscious material. Using my extensive experience of working in this way, I created a course about rhythmic movement training and dream, which I regularly teach in Sweden and Spain.

Both in my courses and in my private practice, I often meet people with pain of the neck, back, or hips due to nonintegrated primitive reflexes. Several years ago, I created a course for physiotherapists and massage therapists in which I focus on exercises releasing muscle tensions that cause pain and osteoarthritis of the back of the neck and the thoracic and lumbar spine. The course also teaches simple exercises to correct a rotated pelvis, normally with a lasting effect. This is one of the most popular courses I teach in Sweden.

Rhythmic movement training in autism spectrum disorder (ASD)

During the last ten years, autism spectrum disorder has become increasingly common and many parents have brought their autistic children to me for rhythmic movement training. The rhythmic exercises have been able to help quite a few of these children to improve, e.g. to develop speech and emotions, but in other children progress was often slow and the exercises caused hyperactivity and severe emotional reactions. The children who have benefited most have been on gluten- and casein-free diets.

I decided to create a course about rhythmic movement training in autism. When I studied the subject, I became convinced that autism is largely caused by environmental factors such as heavy metals, vaccinations, and electromagnetic irradiation, which damage the immune system and the intestines. This causes inflammation of the brain, which explains many of the autistic symptoms. My conclusion was that the rhythmic movements need to be supplemented with other steps (e.g., diet and food supplements) in order to be effective.

The more I studied the environmental causes of the disorder, the more I realized that I had to write a book about it. My book about autism, *Autism: A Disease That Can Heal*, was published in Swedish in 2010. In 2014 it was published in French and Chinese.

Movements that heal

In 2008, I published my book *Movements That Heal* (*Rörelser Som Helar*) in Swedish. It was a summary of what I teach in my courses, supplemented with many case reports to illustrate the development during rhythmic movement training. I translated that book into English and made some revisions, and it was published with Moira Dempsey in 2011. This book is an updated version of the original Swedish book, with more emphasis on environmental causes not only of autism but also of attention and learning problems. The chapter about autism has been completely rewritten, and I have included additional relevant reflexes as well.

CHAPTER 1

How Doctors Traditionally Treat ADHD

Central stimulants and ADHD

Children who are hyperactive, inattentive, easily distracted, quickly tire of what they are doing, or have problems organizing their activities and controlling their impulses are considered to suffer from attention-deficit/hyperactivity disorder (ADHD).

In the United States, there is a long tradition of treating symptoms of hyperactivity with central stimulants. These are highly addictive narcotic substances like Ritalin and amphetamine, which have been used for more than fifty years to treat children with behavioral disorders. During the nineties, the production of Ritalin increased tenfold, and at present, it is estimated that between 7 and 10 percent of American children, mostly boys, are treated with Ritalin or other central stimulants. In addition, during the last years, more adults have started to medicate with central stimulants. In four years, the sale of central stimulants increased from a value of $759 million in 2000 to a value of $3.1 billion in 2004.[1]

The principal promoters of this development have been the big pharmaceutical companies that sell central stimulants and the American National Institute for Mental Health (NIMH). NIMH is a government institution led by psychiatrists who are outspoken promoters of treating hyperactive children with central stimulants.

One of the duties of NIMH is to distribute money for research. According to an article in *U.S. News & World Report*, NIMH has "focused its studies almost exclusively on brain research and on genetic underpinnings of emotional illness ... The decision to reorder the federal research portfolio was both scientific and political."[2]

According to American psychiatrist Peter Breggin, who has criticized the increasing prescription of stimulants to US children, NIMH has granted millions of dollars for research about central stimulants. Nearly all of the money has gone to lifelong ADHD/Ritalin advocates, and none has gone to critics.

Is ADHD a brain dysfunction?

In 1998, NIMH organized a consensus conference with the apparent purpose of getting ADHD recognized as a genetically determined biological disorder. At the conference, a paper was presented that reviewed the whole range of brain scan reports that supposedly showed a biological basis for ADHD. These brain scan studies claimed to have found brain abnormalities in certain areas of the brains of children diagnosed with ADHD. However, in many of these studies, the children diagnosed with ADHD had been treated with stimulants. And none of these studies could be proven to have been based exclusively on children with ADHD who had not been treated with central stimulants.[3] Accordingly, the differences between brains of normal children and children diagnosed with ADHD would more likely be an effect of the medication, which was known to cause brain damage in the areas in question—at least in monkeys.

During the conference, several papers were also presented that highlighted the severe risks and side effects of central stimulants.

After having listened to a number of lectures and studied numerous papers by scientists who had done research on ADHD, the panel cast reasonable doubt on the validity of the ADHD diagnosis.

Most disappointing to the medication advocates of NIMH was the conclusion in the final consensus draft distributed to the press, which stated that there "were no data to indicate that ADHD is due to a brain malfunction."

In 2000, the American Academy of Pediatrics made a similar statement, saying that brain scans and similar studies "do not show reliable differences between children with ADHD and controls."[4]

The MTA study of 1999

Central stimulants have been prescribed to an increasing number of American children for more than fifty years, and many children have taken these drugs for five years or more. In spite of these facts, until recently, none of NIMH's research grants has been aimed at discovering the dangers of longtime use of Ritalin and other stimulants.

In 1999, the so-called MTA study of children who had been medicated with stimulants for one year was published. Until then, most studies followed up treated children at most for a couple of months. One of the principal researchers of this first MTA study, Professor Peter Jensen, made the following statement concerning this study: "We did the best study [that] ever has been done on planet Earth, which helped parents and teachers with these children—and what did it show? It showed that medication still was much more effective for these children."[5] According to well-known British child psychiatrist Eric Taylor, the most important conclusion of the study was that carefully carried-through medication is better than other treatment. This would require medication to be available for children with ADHD.

This first MTA study was a triumph for the pharmaceutical companies and the ADHD/Ritalin advocates among psychiatrists. At the same time, it would turn out to be a disgrace for the child psychiatric community of the world and a catastrophe for the increasing number of children who, as a consequence of the study, would be labeled as suffering from ADHD and treated with stimulants.

The result of this study was widely publicized, causing an ever-increasing labeling of children with ADHD and treatment of them with central stimulants, and this is occurring all over the world. In more than ten countries that I have visited to teach rhythmic movement training, I have heard reports of an increasing number of children who are treated with central stimulants since the beginning of this century.

The recent MTA study, a follow-up study after three years

In 2007, a follow-up of the MTA study, made by the same research team, was published. In this study, the medicated children had been followed for three years.

The result of this study was very disappointing to the research team. One of the principal researchers, Professor William Pelham, made an appearance on the BBC program *Panorama* and stated that after thirty-six months of treatment, there were no positive effects whatever—contrary to what the research team had expected. According to Professor Pelham, nothing indicated that drugs are better than no treatment at all in a longer perspective, and he stressed that this information should be made very clear to parents.[6]

According to Professor Pelham, the report showed that the initial positive effects of the treatment in children with the most severe problems had completely disappeared when the children grew older. The report also established that central stimulants obstruct the normal growth of children, which also affects the growing brain.

The study furthermore demonstrated that central stimulants are connected with more aggressive and antisocial behavior and increased risks of future criminality and drug abuse. Children between eleven and thirteen years old who participated in the study more often used alcohol and illegal substances compared to a control group of classmates. The report concludes that the higher frequency of beginning abuse at an early age requires clinical attention.[7]

With a characteristic English understatement, Professor Pelham said on the BBC program, "I think we exaggerated the positive effects of medication in the first study."

Although the result of this new study was publicized in the United States, the United Kingdom, and Australia, Swedish media remained silent about it. To my knowledge, the findings have not caused any discussion among psychiatrists or in the media anywhere in the world. Medical professionals will seldom, if ever, admit making mistakes, and in the case of central stimulants, psychiatrists seem to behave as if nothing has happened. Perhaps they are waiting for a new study that will contradict the finding of the former one. There's no doubt that the pharmaceutical companies will do anything in their power to produce such a study.

Central stimulants known to obstruct normal growth

If the responsible researchers of the MTA study had done their homework and analyzed the many previous studies on the effect of central stimulants, they would not have been so surprised by the result. Studies in the 1970s already showed that central stimulants reduce growth.

The fact that central stimulants reduce overall growth has been demonstrated in dozens of studies. One obvious cause is that central stimulants suppress appetite, but another cause, one that is more insidious, is that they disrupt growth hormone production. This was shown in 1976 by a Norwegian research group.[8] A study from 1986 of twenty-four young adults who had been treated with central stimulants for hyperactivity in childhood found shrinkage of the brain in more than 50 percent of the cases.[9]

Central stimulants do not improve academic success

In spite of what is often stated by ADHD/Ritalin advocates among psychiatrists, central stimulants do not improve school success in treated children. In 1976, a double-blind study could not show any improvement of scholastic achievement in a group of children treated with central stimulants compared to a control group, even when the treated children's behavior was rated as improved.

On the contrary, the researchers found that central stimulants suppress "desirable behaviors that facilitate learning." In 1992, James Swanson, who is a prominent ADHD/Ritalin advocate, and his colleagues warned, "Cognitive toxicity may occur at commonly prescribed clinical doses." The children become withdrawn and overly focused and may seem "zombie-like." According to Swanson, cognitive toxicity is common and may occur in 40 percent of treated cases, and the overfocusing of attention may impair rather than improve learning.[10]

Increased risk of drug abuse

Also, the increased risk of drug abuse has been demonstrated in earlier research. The US Drug Enforcement Administration (DEA) has repeatedly expressed great concern that treatment with Ritalin will lead to abuse of other drugs. In 1995, the DEA reported, "A number of recent studies, drug abuse cases, and trends among adolescents from various sources indicate that methylphenidate (Ritalin) use may be a risk factor for substance abuse."[11]

At the consensus conference organized by NIMH in 1998, Professor Nadine Lambert of the University of California, Berkeley, presented a unique long-term study comparing future drug abuse in two groups labeled ADHD. The study compared one group who had been prescribed stimulants as children with another group that had not been given medication.

She found a significant correlation between stimulant treatment in childhood and later drug abuse. She told the conference that the prescription of stimulants to children for a year or more was correlated with increased "lifetime use of cocaine and stimulants." She concluded in her paper that childhood use of stimulants "is significantly and pervasively implicated in the uptake of regular smoking, in daily smoking in adulthood, in cocaine dependence, and in lifetime use of cocaine and stimulants."[12]

For obvious reasons, it would be difficult to demonstrate such adverse effects of central stimulants as suppressed growth or risk of drug abuse by following the children only for a year or even less time. Such effects would not become obvious until years later. Many children take central stimulants for five to ten years or more. What will happen to these children in the future can only be speculated on for the time being. No follow-up research for more than three years has been done, and it is doubtful if it ever will be made since the results of such studies are likely to be even more devastating to the pharmaceutical companies than the recently published MTA study.

Why do initially good effects disappear after three years?

If the researchers responsible for the MTA study had investigated previous research on central stimulants with monkeys, they would have been able to predict the outcome that the apparent "good" effects of central stimulants disappear after a few years.

According to a hypothesis endorsed by ADHD/Ritalin promoters, ADHD is caused by insufficient function of the neurotransmitter dopamine in two areas of the brain, the prefrontal cortex and the basal ganglia. The prefrontal cortex is responsible for so-called executive functions: attention, judgment, planning, and impulse control. The basal ganglia, among other things, control our ability to sit still.

The mode of action of central stimulants is to increase the release and prevent the uptake of dopamine in the synapses of the prefrontal cortex and the basal ganglia. As a result, central stimulants will increase the amount of dopamine available in the synapses in these areas, which causes an immediate clinical effect. An overactive child who has been jumping and running around most of the time and been a severe burden to people around him will often, soon after the first dose, be able to sit still and focus on any boring task. This makes an ineffaceable impression on many teachers and parents.

However, this effect is not only short-lived, but it also fetches a high price. The increase of dopamine causes a compensatory dieback of dopamine receptors in the brain, far outlasting the acute drug effect and consequential death of brain cells.[13] In a study on monkeys in

1997, it could be shown that the administration of *two* relatively small acute doses of amphetamine (2 mg/kg, four hours apart) produced persistent, marked decrease in dopamine synthesis and concentration up to three months later. One animal showed continued dysfunction even after eight months.[14]

In children who medicate with amphetamine and other stimulants, drug levels on a milligram/kilogram basis can be as high as those reported to have caused brain damage in animals in several studies.

According to the MTA study, central stimulants lose their positive effect when children are treated for more than a year up until three years. This is logical considering the long-term effect of central stimulants on the brain, which is to decrease dopamine-producing nerve cells. Consequently, the dose of central stimulants must be increased to produce the same effect, and in the end, central stimulants will have no positive effect at all due to extensive loss of dopamine-producing brain cells.

A long-term Australian study confirms the MTA study

In the beginning of 2010, the West Australian Department of Health published a *long-term outcome study* [15] on the use of stimulant drugs for treatment of ADHD.

The data used was from 131 patients who were tracked for twenty years by the department. These patients were matched to a similar group with the diagnosis of ADHD that had not been treated with central stimulants.

The study found stimulant medication increased blood pressure, had nil results on academic performance, and didn't improve behavior. For those on medication, it is ten times as likely that classroom performance will be below average. The physical effects of the medication follow the child into adulthood.

In an interview on Australian radio, coauthor Professor Lou Landau expressed dismay about these outcomes because these contradict many published industry-sponsored short-term studies. The authors concluded the following: "The lack of significant improvements in long-term social, emotional and academic functioning associated with the use of stimulant medication suggests a purpose-designed, longitudinal research study should be conducted to better understand the suspected long-term social, emotional and educational benefits of stimulant medication in the treatment of ADHD."

Concerta not approved to treat adults in Europe

In 2010, the pharmaceutical company Janssen-Cilag submitted an application to get Concerta approved for adults in Europe. However, Janssen-Cilag withdrew its application after the European Medicines Agency (EMEA) found that Concerta was not acceptable for adults and had a "negative benefit/risk balance." It was not more effective than a placebo, and serious safety problems were found, including a considerable abuse potential and risk for aggression, anxiety, and depression.

The development in Sweden

Between 2000 and 2011, there was a tenfold increase of children who were prescribed central stimulants, from two thousand children to nearly twenty-five thousand in 2011. This is a remarkable development considering the fact that Ritalin was withdrawn from the marketplace in 1968 due to its great popularity causing widespread abuse. In the 1970s and 1980s, central stimulants were seldom prescribed to children. They could only be prescribed with a special license from the National Board of Health.

In the end of the 1990s, the number of children who were prescribed central stimulants started to rise considerably. Leading child psychiatrists estimated that around ten thousand children with ADHD would eventually need treatment with these drugs. But in 2010, already twenty thousand were being treated and in 2013, more than fifty thousand.

The Swedish National Board of Health recommends stimulants

In 2004, the Swedish National Board of Health published a booklet titled "Briefly about ADHD in Children and Adults."[16]

The board emphasized heredity as the cause of ADHD and wrote the following: "Heredity is brought about by genes. Genes control the transmitter substances that convey information between the neurons of the brain. Shortage or insufficient effect of these substances in certain areas of the brain causes changes of psychological/cognitive function that may cause problems for the child to control his/her behavior. This in turn will result in the typical symptoms of ADHD, like restlessness, attention problems, and impulsivity."

This sensational statement about the cause of ADHD did not have any scientific foundation and was contrary to the research presented at the American consensus conference organized by NIMH in 1998 and the final consensus draft of this conference. The board recommended the use of central stimulants as a treatment of ADHD and emphasized how well documented these drugs are, especially in large studies on children, how effective they are, and how slight their side effects are. According to the board, there are no other psychoactive drugs that have been so thoroughly studied as central stimulants, and it makes the following statement: "Due to the rapidly growing knowledge about ADHD in Sweden and the fact that we now have taken part in international experiences of medication, the number of children treated with central stimulants has rapidly increased in the same way as in other countries."

The National Board of Health praised central stimulants for increasing the power of concentration and decreasing hyperactivity. Moreover, "the drugs seem to improve cognitive abilities like problem solving."

Concerning the danger of addiction and future abuse, the National Board of Health stated there is no such risk and claims that treatment with central stimulants "instead seems to diminish the risk of future abuse."

The board also wrote that follow-up studies of children diagnosed with ADHD often show a gloomy picture, "with poor academic and professional success and often serious and psychiatric problems in adult age." The Swedish National Board of Health considered that ADHD should be met as a public problem "since it concerns many individuals and has a serious effect on their health, development, and possibilities to live ... fully adequate [lives] as adults."

The new MTA study refutes the board's standpoints

The recent MTA study has shown that after three years, medication stimulants have no positive effect whatever. They are simply no better than no treatment at all. Therefore, it must have been a mistake to label stimulants as effective drugs for treating ADHD. And contrary to the claim of the Swedish National Board of Health that central stimulants "seem to diminish the risk of future abuse" the MTA study has demonstrated that they cause an increased risk of future drug and criminality.

Because of this finding, the recent MTA study has also corroborated that it is not the diagnosis of ADHD but the medication with central stimulants that has "a serious effect on children's health, development, and possibilities to live ... fully adequate [lives] as adults."

Now that the board knows that stimulants are not at all effective after three years' use, one would expect it to come out and correct its previous statement that it is still not known how effective central stimulants are after many years of treatment. One would also expect it to follow the advice of one of the principal researchers of the MTA study and make it clear to parents that nothing indicates that drugs are better than no treatment at all in a longer perspective.

However, the board has not done this. After seven years of knowing the effects, it has still not commented on the MTA study at all. Nor does it seem likely that it ever will. And what could be the reasons for that?

Harald Blomberg, MD

The reasons for the cover-up of the recent MTA study

Swedish psychiatrists and child psychiatrists have been most successful in persuading politicians to give resources to psychiatrists to label children and adults with ADHD and treat them with central stimulants. In addition, psychologists and social workers have been employed and have received special training to be able to assist doctors in diagnosing persons who are supposed to need treatment with stimulants. In the beginning, there was some resistance to drugging children with stimulants in some parts of Sweden, but with the united forces of the media and the National Board of Health, the drugging enterprise was soon riding on the crest of a wave, creating a snowballing effect: When more children were diagnosed and treated, more doctors, psychologists, and social workers were needed, causing more children to be diagnosed and so forth. By now, there is a host of professionals to serve the ever-increasing undertaking of diagnosing and treating ADHD.

Considering these circumstances, it is becoming obvious that neither the National Board of Health nor doctors, psychologists, or other health workers, nor politicians, have any interest in revealing the true facts about central stimulants and ADHD.

The National Board of Health and its psychiatric experts naturally do not want to lose their credibility by having their incompetence exposed. Doctors and health workers also want to protect their jobs and livelihoods. Politicians would not want to be exposed as swallowing information without proper study and therefore wasting taxpayers' money on a scam.

So Swedish taxpayers will have to continue paying. And the drugging of Swedish children will continue without restriction and without any positive effects in the end. But undoubtedly, these drugs will cause a lot of damage in children, which in the individual case will increase the longer the drugging is permitted to continue.

CHAPTER 2

AN ALTERNATIVE WAY OF REGARDING AND TREATING ADHD

Normal small children have symptoms of ADHD

By studying our smallest children, it is possible to get other angles of approaching attention deficit disorder and how it can be remedied rather than those advocated by psychiatric establishments and the pharmaceutical companies. It is normal for children around the age of one year to have a similar behavior to that of children with ADHD if they are allowed to move around freely and not forced to sit in baby seats or car seats for long periods.

They move around, cling and climb, and have problems sitting still. They are impulsive, easily distracted, and quickly tire of what they are doing. They have difficulties listening, following instructions, and organizing their activities. They have problems controlling their emotions and temper.

However, unlike older children who are labeled as suffering from ADHD, normal children, all by themselves, manage to overcome their attention disorder and hyperactivity as they grow older. How are children who develop normally different from children who develop ADHD? What secret knowledge, unknown to the experts, do they have that enables them to prevail over their attention problems?

Could it be true that children with ADHD have a genetically caused shortage of transmitter substances in the brain—as the

Swedish National Board of Health claims—or can there be other more plausible and less far-fetched explanations?

The plastic brain

The British obstetrician and researcher Robert Winton has written popular books and made popular television series. In his book *The Human Mind*, he describes the ability of the brain to regenerate itself: the plastic brain.

In the 1940s, brain researchers had already found that the communication between neurons goes in both directions. When a nerve cell of the brain—a neuron—receives a signal from another neuron, it passes on the signal to other neurons and at the same time sends a feedback signal to the cell that originally sent the signal.

When this first process is over, another process begins. The affected neurons on their part multiply the connections with each other by growing new nerve synapses and increasing the supply of transmitter substances thereby facilitating the firing of the nerve signals in the new pattern.

"This feedback and learning mechanism of the brain signifies that every nerve connection of the brain is informed of to what extent it has contributed to the end result of the communication ... next time to create new nerve connections and release more transmitter substances."[17]

The infant's brain is undeveloped

The brain of the infant is very immature. In the newborn baby, only the brain stem functions properly while the other parts are made use of only to a small extent. Before an individual is able to bring into play all of its brain, the nets between the nerve cells of the brain must develop by growth of branches from the nerve cells and the nerve fibers must develop an isolating sheath of myelin. This maturing of the brain will take place throughout childhood; however, it is the very first year that is the most crucial period for laying the foundation for later development. It has been estimated that with every minute in the life of a newborn baby, more than four million new nerve cell branches are created in the brain of that infant.

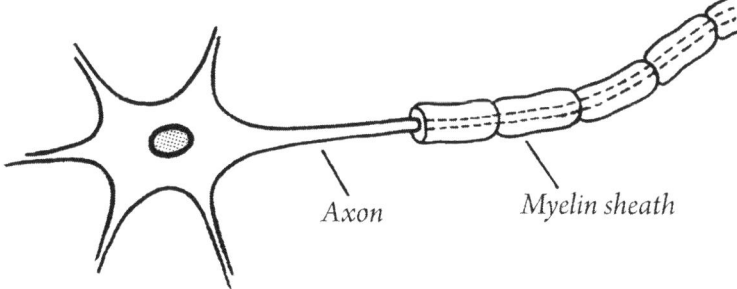

Fig. 1 Nerve cell with axon and myelin sheath

This process does not happen by itself. The brain needs stimulation from the senses for the branching off and the myelination to occur, especially stimulation from the vestibular, the tactile, and the kinesthetic senses. The baby gets such stimulation by being touched, rocked by its parents, and by continually making rhythmic baby movements on its own. Such movements develop in a certain order according to an inborn program with individual variations. Turning around, crawling on the stomach, rocking on hands and knees, and crawling on hands and knees are some important milestones in this development. The stimulation the brain of the infant gets from such rhythmical baby movements during the first year of life is fundamental for the future development and maturing of the brain.

When the nerve connections and synapses of the brain increase in number, additional parts of the brain start to perform their functions in new nerve patterns stimulated by the child's movements. As described above, this process continues automatically even when the nerve cells do not get direct stimulation. At the same time, such nerve connections corresponding to old behavioral patterns that the child no longer needs are pruned.

In children who have not had sufficient stimulation of this kind, the maturing of the brain is delayed or impaired. Such delayed development can appear as an attention disorder, with or without hyperactivity.

The triune brain

American scientist Paul MacLean has studied the development of the brain in reptiles, mammals, and humans. According to him, the human brain consists of three layers that surround the brain stem, which roughly corresponds to the brain of the fish. The brain stem is part of the neural chassis, which also includes the spinal cord. The three layers of the triune brain surround the brain stem like layers of an onion.[18]

According to MacLean, the neural chassis might be likened to a vehicle without a driver and the three layers surrounding the brain stem can be likened to three guiding operators, each one with its own intellect, its own memory bank, and other functions.

Next to the brain stem is the reptilian brain, or R-complex, which corresponds to the new parts of the brain that reptiles developed. In humans, the reptilian brain is called the basal ganglia, one task of which is to control our postural reflexes (i.e., our ability to stand, walk, and keep our balance).

The reptilian brain must also inhibit the primitive reflexes, which are inborn, stereotyped movement patterns controlled by the brain stem. These primitive reflexes constitute the movements of the fetus and the newborn infant and must be transformed into the postural reflexes in order for the child to be able to rise, walk, and keep her balance. The basal ganglia also regulate the level of activity of the child to ensure that the child is not revved up most of the time.

Outside the reptilian brain is the mammalian brain, or the limbic system, which controls our emotions, memory, learning, and play, among other things.

On the outside is the neocortex or the human brain. Signals from the sense organs must reach the neocortex and be processed there in order for us to be aware of what happens around us and to be able to act consciously. The very front part of the neocortex, the prefrontal cortex, is of crucial importance for our judgment, attention, power of initiative, and control of impulses.

Rhythmic baby movements necessary for the linking up of the brain

When we are born, all parts of the triune brain have been set up, but are not yet working properly. In order for all the parts of the brain to function as a unit, they must be developed and linked up to each other. This is achieved by the rhythmic infant movements that stimulate the production of new neurons, the growth and branching off of the neurons, and the myelination of the nerve fibers.

The infant needs to have sufficient muscle tone in order to be able to move around and stimulate the linking up of the different parts of the brain. Also, for sufficient muscle tone to be established, the infant must be touched, hugged, rocked, and allowed to move around freely. Such stimulation sends signals from the sense organs of the tactile, vestibular, and kinesthetic senses to those centers of the brain stem, the task of which is to regulate muscle tone. If the baby gets insufficient stimulation from these senses, the muscle tone of the extensor muscles will be low.[19] This may cause difficulties for the baby in being able to lift his head and chest and move around, further reducing the stimulation from the vestibular, tactile, and kinesthetic senses, leading to a vicious circle.

When the baby is unable to move around freely, too little stimulation is conveyed to the neocortex via the reticular activation system (RAS) of the brain stem. The task of this system is to arouse the neocortex. When there is insufficient arousal of the neocortex, the child will become sluggish and inattentive to sensory signals. Moreover, the nerve cells and the nerve nets of the neocortex will not develop properly.

The cerebellum is also important for the linking up of the brain and for the power of attention. The task of the cerebellum is to make our movements rhythmic, coordinated, and smooth. From the cerebellum, there are important nerve connections to the prefrontal cortex and the centers of speech in the left hemisphere.

At birth, the cerebellum is undeveloped and it grows substantially after the age of six months. The rhythmic baby movements develop the nerve nets and nerve cells of the cerebellum and its connections to the frontal lobes. That is one reason why the rhythmic baby movements are so important for the linking up of the frontal cortex and the development of the power of attention and speech.

Why babies have challenges sitting still and being attentive

The fact that the nerve nets of the brain have not developed and the different levels of the brain have not been linked up explains why infants do not behave like small adults. Infants are not good at maintaining their attention, concentrating on a specific task, or controlling their impulses because the nerve nets of the neocortex, especially the frontal lobes, have not developed.

Infants have problems regulating their level of activity and normally, by the age of ten to twelve months, they move around most of the time and find it difficult to sit still. Since the basal ganglia have not yet developed properly and are not linked up with other levels of the brain, most normal babies are hyperactive at this age.

On the other hand, babies who are not able to move around sufficiently, due to low muscle tone or other circumstances, get too little stimulation of their neocortex and frontal lobes and will become sluggish, hypoactive, inattentive, and late to develop.

Similarities between small children and children with ADHD

It has been shown that attention deficit problems and hyperactivity are common features in small children and children who have been identified as suffering from ADHD. In both groups, there are many signs that the basal ganglia do not function properly; these signs are difficulties in regulating the level of activity, active primitive reflexes, and balance problems.

It is also common that children with an attention disorder have an inability to make simple movements in a rhythmical and smooth way, which indicates that the nerve nets of the cerebellum have not been properly developed. Since the cerebellum is of crucial importance for the proper functioning of the prefrontal cortex, this inability can be an important contributing factor behind attention problems and impulsivity.

Many children with ADHD have low muscle tone and hunched up posture, which causes shallow breathing and insufficient arousal of the neocortex. Such children may alternate between hyperactivity and passivity, the hyperactivity being a way to stimulate the neocortex by moving around.

Attention disorder as delayed maturity of the brain

As we have seen, experts consider ADHD to be genetically caused. An alternative explanation would be that a delayed or thwarted maturity process of the brain causes the symptoms. For some reason, the brain of the child has not received sufficient stimulation for the neurons to branch off and create new synapses. Lack of stimulation may also obstruct the formation of myelin around the nerve fibers. If there is insufficient myelination of the nerves, it will affect the speed of propagation of the nerve signals. Taken together, all this may obstruct the development of the different parts of the brain and their linking up to each other, preventing the brain to function well as a whole.

Everything that obstructs the motor development of the child and prevents her movements will therefore obstruct the development of the brain.

Many circumstances such as prematurity, brain injury inflicted during delivery, hereditary factors, vaccinations, electromagnetic fields and microwaves from cell phones, food intolerance, toxicity, or disease may affect the motor development of the child. Such factors may cause the infant to omit important steps of his motor development, in that way hampering his motor development and the maturing of his brain.

A lack of stimulation from those closest to the child, being left alone without tactile or vestibular stimulation, or being forced to spend his time in baby seats instead of moving around on the floor will also prevent the brain from maturing properly.

Rhythmic movement training

As shown, there are many similarities between infants and children with ADHD regarding behavior and immaturity of the brain.

One must therefore ask if children with ADHD or ADD can improve by imitating the rhythmic movements that infants spontaneously make. Such movement training has been used in Sweden for more than twenty-five years.

Rhythmic movement training was developed by Kerstin Linde and is founded on the natural rhythmic movements of the infant. To be effective, these exercises need to be done daily for ten to fifteen minutes. The movements are done in lying or sitting position or on hands and knees.

The movements used in the training are active or passive rhythmic whole body movements. Passive movements can occur by pushing the client's body rhythmically from the feet with the client in a supine position or from his hip in the direction of the head in a fetal position. With the client lying on his face, the bottom can be rhythmically pushed from side to side. In a supine position, the legs may be passively rolled so that the big toes meet in the middle.

These movements can also be made actively. In a supine position with knees bent, the client may push from the feet rhythmically or roll his legs back and forth from the sides to the middle so that the big toes meet. In a prone position, the client may roll his bottom from side to side. Other active movements include rocking on all fours and crawling.

Whether done actively or passively, these exercises are suitable for every client, no matter how disabled. Ideally, these movements should be made in a very exact way. In severely handicapped people, this is of course impossible, and one long-term goal is to teach the clients to do them more precisely.

One can easily see that these movements give a strong stimulation to several senses. The movements of the head stimulate the vestibular sense. The rhythmical pushing along the spine from the feet or bottom stimulates proprioception in many joints and the inner organs. The rhythmic movements also stimulate the tactile sense organs of the skin by the friction between the back or front and the floor.

The effects of the rhythmic movements on behavior

The sensory stimulation caused by the rhythmic movements stimulates the nerve nets of the brain stem, cerebellum, basal ganglia, and neocortex to develop. This causes attention and concentration to improve and hyperactivity and impulsivity to decrease.

Signals from...
- *Sense of Balance*
- *Tactile Sense*
- *Joint- & Muscular Sense*

→ BRAIN STEM →

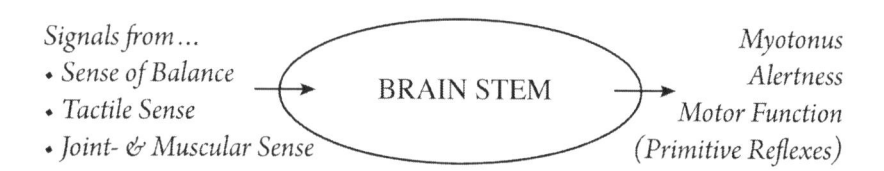

Myotonus
Alertness
Motor Function
(Primitive Reflexes)

Fig. 2 The effect of the rhythmic infant movements and rhythmic movement training

The rhythmic movements also increase the muscle tone of the extensor muscles that straighten the back and keep the head in an upright position. Body posture, breathing, and endurance will improve, and the neocortex will be aroused by stimulation via the brain stem, which will improve attention and concentration.

The rhythmic movement training will stimulate the cerebellum and its nerve paths to the prefrontal cortex, which will also improve attention and concentration and diminish impulsivity.

The rhythmic training also stimulates the basal ganglia to mature and integrate the primitive reflexes, which will facilitate the ability of the child to regulate its level of activity and be still.

The significance of the rhythm in rhythmic movement training

The spontaneous rhythmic infant movements stimulate, organize, and develop the brain of the baby by nerve signals from the senses and by weak electromagnetic frequencies.

According to the biologist James Oshman, the body is a living matrix in which all parts are in contact with each other, from the skin to the cell nuclei. The information between the different parts is transmitted not only by nerve signals but also by electromagnetic impulses of different frequencies.

The rhythmic movements cause an intermittent, ever-changing stimulation of the brain from the vestibular, tactile, and proprioceptive senses. In this process, the nerve signals are transmitted by transmitter substances (e.g., dopamine, glutamate, and GABA). Such changing, oscillating stimulations are much more effective than continuous ones since continuous or similar signals cause a decreased response to the stimulus, which is referred to as habituation.

Information throughout the living matrix of the body is also transmitted by different kinds of energy. Energy is fundamentally vibration in the form of light or electromagnetic energy, sound, chemical, or mechanical elastic energy. Many of the most important molecules of the muscles and skeleton are formed as spirals. This makes them elastic and gives them good resonance properties. When the body is set in oscillation by the rhythmic movements, weak electromagnetic fields are created, which transmit information to all parts of the body, especially to the nervous system and the brain.

How to develop a training program

In order to know what exercises a child with problems needs to do, one must get an idea about the difficulties of the child. It is important to conduct a thorough interview—for instance, using a questionnaire—to have an understanding of the anamnesis and the problems that the child feels most affected by. The interview should be supplemented with an examination of the child's motor abilities. It is especially important to find out if the child has problems doing simple rhythmic exercises, and primitive reflexes should be assessed. In learning problems, a simple examination of vision is very helpful. Based on the interview and the result of the examination, it is possible to get an idea about which functions the child needs to improve.

Children with symptoms of ADHD and learning difficulties have always retained primitive reflexes. However, they do not need to have obvious motor problems. Sometimes children with attention and learning difficulties may even have good motor abilities and be good at sports and gymnastics. More often, children with attention problems have more prominent motor problems such as low muscle tone, poor posture, and difficulties doing simple rhythmic exercises.

The training program of the child should be based on the interview and the result of the motor examination. At least in the beginning, this program should not take more than ten minutes to do, and it should be done daily, no less than five times a week. As the motor abilities improve and the reflexes are integrated, the program should be changed. Concurrently, with the integration of the reflexes and the improvement of the rhythm in the simple rhythmical movements, the symptoms of the child will improve. Attention, control of impulses, and the ability to sit still are some of the symptoms that usually improve. In most cases, it will take a year or more before the symptoms are permanently gone. If the exercises are not done for a sufficiently long time, the new pathway of the brain may not have time to consolidate and some of the symptoms may reappear.

Many children and adults who have retained primitive reflexes have never had any attention or learning problems. They may instead have other problems such as visual, motor, or emotional problems, or pain of the muscles and joints.

Case report: Anna

A case report may illustrate the effects of RMT in attention disorder. It shows how the rhythmic movements improve attention and diminish impulsivity and hyperactivity.

Anna was ten years old when she began rhythmic movement training.

Her motor development had been normal except for fine motor development. She crawled on hands and knees as a baby and walked at the age of one year. She had great problems concentrating and sitting still at school. She was easily distracted and had poor perseverance. She had no problems reading and writing but had great problems with mathematics. She had her own assistant during mathematic lectures, and if the assistant was absent, she would do nothing but run around disturbing the other students.

She acted on impulse and had great problems paying attention and following instructions, especially during physical education which she did not want to attend. Her ankles were weak, and she easily sprained them. She had great fine motor problems, especially tying shoelaces and doing up buttons. Her handwriting was bad.

Anna had emotional problems. She was afraid of the dark, anxious and apprehensive, especially at night. She had severe difficulties in her peer relations. The girls in her class used to tease her, and she would run away and hide.

At her first visit, I tested for her primitive reflexes, many of which were active. The spinal galant reflex was especially active, explaining her inability to sit still and to wear tight clothes. Her Moro reflex was also extremely active, causing sensitivity to sounds and touch and many of her emotional problems.

Her Babkin and grasp reflexes were also active, causing problems with fine motor ability.

Anna's training

Anna visited me about once a month for a little more than a year. She did rhythmic exercises at home for ten to fifteen minutes every day. During her visits, she got new rhythmic exercises to do. Some of the rhythmic exercises she continued to do most of the time. She got assistance in order to do learn to do them correctly. In addition, she got special exercises for reflex integration, which her mother helped her with.

After four months, her mother noticed that she had become more defiant and ill tempered than before. After five months, these symptoms had decreased and she was more self-confident. She could concentrate better at school, and schoolwork, including mathematics, went better. A few weeks later, she had caught up with her classmates in mathematics. Also, physical education went better and she liked to attend. She especially liked long-jumping and high-jumping.

After half a year, she changed to another school after summer vacation, and her mother chose to say nothing about her problems. It turned out that she could concentrate well and no longer needed any assistant. She had no problems in her relations with her classmates, and she was no longer teased. Her ankles had become much stronger, and she now started soccer training. Her fine motor ability had improved considerably.

After a little more than a year of movement training, she was no longer afraid of the dark. She was no longer hypersensitive to sound and touch and not so easily disturbed. She had no problem sitting still. Her fine motor ability had improved considerably, although she still had some problems doing up buttons. She had practically no difficulties with concentration, and her endurance was good.

CHAPTER 3

Two Different Ways of Looking at Children with Challenges

Diseases and their causes are cultural phenomena

As we have seen, official scientific experts, whether they are doctors, psychologists, or dyslexia researchers, are in the habit of labeling children with challenges as suffering from specific diseases or dysfunctions such as ADD, ADHD, autism, dyslexia, and so forth. Such diagnoses can be regarded as social constructions and cultural phenomena. Different subjective symptoms are brought together to form disease entities to be discerned and treated by doctors and other experts.

In order to treat a disease successfully, as it has been diagnosed, it is advantageous to know what caused it. After medical experts have identified and characterized a disease, which seems to be the easy part, they start looking for the cause. In general, scientific medicine looks for a solitary cause of any disease or diagnosis. Especially the discovery that infectious diseases were caused by bacteria promoted the notion that each disease had a solitary cause. However, different kinds of causes have been popular in different epochs. After bacteria, viruses became popular as agents that cause diseases, and now the genes are the most popular causes of diseases among the medical profession. Obviously, not only diagnoses but also the recognized causes of diseases and dysfunctions are cultural phenomena, which may be illustrated by the development in the psychiatric field.

The triumph of the medical model in psychiatry in the 1980s and 1990s

When I trained as a doctor in the 1960s, psychological and emotional factors were considered to be relevant as causes of disease, especially of mental diseases. Psychologists and social workers, counselors, family therapists and other non-psychiatrists were taking over the mental health field. As upholders of the traditional medical paradigm that the mind is only an effect of physical processes, psychiatrists were becoming marginalized. In the beginning of the 1980s, the American Psychiatric Association decided to create a partnership with drug companies, which would give psychiatry access to funds from the drug companies to promote the medical model, psychopharmacology, and the authority and influence of psychiatry.[20] This turned out to be a very successful strategy, if not for the patients, at least for the psychiatrists and pharmaceutical companies.

This process was not limited to the United States but spread all over the world. In Sweden, leading psychiatrists have been especially successful in promoting the medical model and pharmacological treatment of behavioral symptoms in children. In a way, Swedish child psychiatrists seem to have taken the lead in the promotion of the medical model in their specialties. While there is still no consensus among American pediatricians and child psychiatrists that ADHD is caused by a brain malfunction, the Swedish National Board of Health boldly states that the typical symptoms of ADHD, like restlessness, attention problems, and impulsivity are due to a genetically determined lack of transmitter substances. In addition, Swedish psychiatrists also have managed to convert Swedish psychologists and social workers to the medical model and make them their henchmen in labeling and psychopharmacological treatment of children.

Harald Blomberg, MD

The disreputable connection between race biology and genetics in Sweden

The willingness with which Swedish psychiatrists have enthusiastically picked up the genetic model of explanation for ADHD may seem odd. However, I believe there are historical and cultural reasons for the attraction of Swedish psychiatrists to the genetic model and the absence of questioning of the same.

In the 1930s and the beginning of the 1940s, race biology was a generally accepted ideology in Sweden, as in other countries. The politicians had funded an institute of race biology in Uppsala, which took on the task of studying and documenting the different race groups in Sweden. By measuring the skulls and photographing the naked bodies of the Lappish and Finnish sections of populations in the north of Sweden, the researchers considered themselves able to prove scientifically the racial inferiority of these groups compared to the dominant so-called Aryan Swedish population. Many people in Lapland can still tell how they as children were forced to take part in these studies against their will. They are quite conscious of the fact that if the Nazis had won the war, many of them would have been brought to extinction camps.

After the Second World War, race biology became obsolete and the Institute of Race Biology in Uppsala was not allowed to continue its activity. It was quietly merged with the Department of Genetics of the faculty of medicine, which still harbors its research material. When I studied genetics in Uppsala during my medical training in the 1960s, the history of the department was discussed among the students but never openly with the teachers. Nor in the following years has there been any ideological settling of accounts of race biology among doctors and geneticists, nor has the topic been adequately studied by historicists and sociologists. Unlike the aborigines of Australia, the Lappish and Finnish sections of the population in the north of Sweden are still waiting for an official apology for the discrimination and racist treatment of them by the state.

I believe that this absence of settling the accounts of the official Swedish racist ideology still influences the attitudes and actions not only of the Swedish medical profession but also of many sectors of Swedish society. The total absence of questioning that ADHD is a genetically determined lack of transmitter substances in the brain is especially significant in this respect. So is the readiness of the medical profession to treat the labeled children with addictive and seriously harmful drugs as central stimulants. So is the support of these practices by the media and the benevolent funding of it by the politicians. As has been demonstrated, the idea that ADHD is a genetically determined lack of transmitter substances in the brain has no more scientific foundation than the idea that the Lappish and Finnish sections of the Swedish population were genetically inferior, and it is furthermore contradicted by many facts.

Eight of ten Swedish children diagnosed with ADHD come from lower social groups

In Sweden, Christoffer Gillberg, professor of child psychiatry in Gothenburg, has studied children labeled as suffering from ADHD. He is one of the experts who have promoted the view that ADHD is a genetically determined biological disorder. When sociologist Eva Kärve analyzed his research material, she found, to her surprise, that eight out of ten children who had been labeled ADHD belonged to lower social groups. There was no evident explanation of this fact. To scientists who believe that ADHD is genetically determined, the obvious conclusion of this finding would be that lower social groups are genetically predisposed to have a higher incidence of ADHD than the higher social groups. In other words, they are genetically inferior. However, such an idea is not politically correct in Sweden and is not embraced by the scientific community; therefore, the National Board of Health cannot expound it. On the other hand, if the genetic predisposition of ADHD is the same in all social groups (as is the accepted notion), eight out of ten children with ADHD cannot possibly belong to the lower social groups. Such a result would refute the idea that ADHD is a genetically determined biological disorder. In the same context, Gillberg's result would point to social and environmental circumstances as causal factors of ADHD.

However, if it were a generally embraced concept among scientists that the lower social strata were genetically predisposed to develop ADHD, Gillberg's result would not only give scientific corroboration of their genetic inferiority but also of the notion that ADHD is a genetically determined condition.

Treating children with central stimulants is a violation

The efficiency and thoroughness with which Swedish politicians and the medical profession have organized the diagnosing and treatment of children with ADHD does not seem to be inferior to the effectiveness with which the Institute of Race Biology organized the measuring of the skulls of the Lappish and Finnish sections of the population in Sweden in the 1930s and 1940s. It would even be admirable if the purpose and consequences were not so destructive.

Like the Lappish and Finnish children when their skulls were measured in the 1930s, children being diagnosed with ADHD have no say in the matter. They are prescribed central stimulants whether they want them or not. Should they initially protest, the drug would soon make them compliant and more willing to do as they are told.

Unlike the Lappish and Finnish children of the 1930s and 1940s, children with ADHD are not threatened with physical extinction—only their brain cells are. As stated, many animal studies on monkeys have shown that relatively low doses of central stimulants cause destruction of brain cells and permanent brain damage, especially in the frontal lobes and the basal ganglia. When the central stimulants cause the synapses to be overstimulated by dopamine, they will perish, causing impaired function of these parts of the brain. In children treated with stimulants, drug levels on a milligram/kilogram basis can be as high as those reported to cause brain damage in animals, according to a scientific report.

Stimulants make children manageable at the cost of their spirits

The promoters of central stimulants (e.g., the Swedish Board of Health) emphasize how these drugs improve the functioning of children with ADHD and facilitate the intercourse with other children.

What really happens to children treated with central stimulants is, however, something quite different. They become more compliant and more willing and able to do as they are told, especially regarding boring and repetitive tasks at school. In addition, they become less spontaneous and curious, and they often withdraw from contact and play with other children.

Numerous animal studies of central stimulants show striking similarities with the reactions of children. Peter Breggin summed up the result of these studies in the following way at the NIMH conference: "First, stimulants suppress normal spontaneous or self-regulated activity, including curiosity, socializing, and play. Second, stimulants promote stereotyped, obsessive-compulsive, overly focused behaviors that are often repetitive and meaningless."[21]

Even if such reactions may result in less conflicts with those around and in that way facilitate intercourse with others, one must question if they really imply that the functioning of the children has improved. Should not such conduct instead be a reason for concern and require careful consideration?

In the United States, a consortium of attorneys is bringing a series of class action suits against Novartis (the manufacturer of Ritalin) and the American Psychiatric Association.

One of these attorneys writes in the forward of Peter Breggin's book *Talking Back to Ritalin*: "I asked myself if the large sums of money earned by the pharmaceutical industry could corrupt their research in the same way as in the tobacco industry. Much as the tobacco industry promoted and marketed its products with children in mind, I began to wonder if our vulnerable children were again being targeted for corporate profit. Ultimately, stimulants steal childhood. They make children more manageable at the cost of their [spirits]."

Side effects of central stimulants

According to promoters of central stimulants (e.g., the Swedish Board of Health), these drugs have relatively few side effects. However, the real facts are quite different.

From Breggin's survey, it is evident that the side effects of stimulants, far from being insignificant, are both serious and extremely common. In several studies, the frequency of side effects is more than 50 percent. The most common side effects are loss of appetite, drowsiness, withdrawal, loss of interest in others, and depression. In one study of forty-one children between four and six years of age, 75 percent suffered from loss of appetite, 62 percent from drowsiness, and 62 percent were uninterested in others. In another study of eighty-three somewhat older children, 45 percent had side effects that mostly included withdrawal, sadness, or crying.

Obsessive-compulsive symptoms are very common side effects. They appear as a compulsive repetition of simple activities like endlessly playing games on the computer. In one study of forty-five children, 51 percent developed compulsive symptoms that in certain cases were very serious. One child became so obsessed with doing a good job raking leaves that he would wait for each one to fall from the tree. Another played with Legos for thirty-six hours without breaking to eat or sleep.[22]

In another study, 42 percent of the children produced an obsessive overfocusing after a single dose of stimulants. The children were sometimes unable to stop performing tasks that had been assigned to them.

Tics and movement disorder are also common side effects. In a study with forty-five children, 58 percent developed tics and abnormal movements. In another study with 122 children, 9 percent developed tics and abnormal movements. One child did not recover and developed an irreversible syndrome with facial twitching, head turning, lip smacking, forehead wiping, and vocalizations.[23]

A Canadian study from 1999 could demonstrate that at least 9 percent, and probably a bigger share of the ninety-eight children who had been treated with central stimulants, developed psychotic symptoms.

It has long been known that central stimulants decrease the blood circulation of the brain, damage the blood vessels, and cause hemorrhages. In a textbook of psychiatry, Jaffe writes, "In monkeys, the toxic effects of chronic amphetamine use include damage to the cerebral blood vessels, neuronal loss (brain cell death), and micro haemorrhages."[24]

Not until recently has attention been paid to such side effects in children and adults after many reports of rise of blood pressure, stroke, and sudden heart death in treated patients. In 2005, the central stimulant Adderall was temporarily withdrawn from the market in Canada after reports of twenty cases of heart attacks and twelve cases of stroke.[25] In February 2006, Reuters reported that fifty-one patients who had been medicated with central stimulants had suddenly passed away, which caused the FDA to urge prescribers to pay attention to heart attacks and high blood pressure. Another thirty deaths in patients who had been treated with Ritalin had also been reported.

Strattera

Strattera is produced by the drug company Eli Lilly and was originally an antidepressant drug that turned out to have no effect on depression. It was instead introduced in 2002 as a remedy for ADHD. It is not addictive like other drugs used in ADHD. However, its effects are questionable and its side effects are appalling. According to a Swedish study of one hundred children medicated with Strattera, the children experienced no positive effects of the drug. However, many children complained of one or several negative side effects. Stomach pain, headache, tiredness, loss of appetite, and nausea were the most common, with an incidence of about 40 percent. Around 10 percent of the children had psychological side effects like irritation and depression.[26]

By 2005, Eli Lilly had already received reports about 10,998 psychiatric reactions.[27] The same year, the European Medical Products Agency had warned about Strattera causing hostility and emotional instability in children, and there were also international warnings that Strattera increased the risk of suicide.

In 2006, the FDA made a survey of all reports of negative psychological side effects according to which 992 cases of aggression or violent behavior and 360 cases of psychosis had been reported. In 90 percent of these cases, there was no previous history of similar symptoms.[28] It should be emphasized that at the very most, 10 percent of negative drug reactions are reported.

A new investigation of cases reported to the FDA (2004–2007) with death as an outcome showed that thirty-one children and teenagers had died in the United States while under Strattera treatment. Nineteen of these committed suicide. In addition, six children and teenagers were reported to have died in Europe. During the same period, thirty-seven adults had died in the United States and in Europe. Seventeen of them committed suicide. Altogether, seventy-eight persons treated with Strattera died within three years.

Increasing prescription in spite of serious side effects

Central stimulants are not the only drugs that cause serious and life-threatening side effects and deaths of many patients. Such effects of drugs are common, and the medical profession tolerates them on the condition that more patients are believed to survive with drug treatment than without such treatment. Also, if the drug prolongs the life of the patient, a certain mortality is tolerated.

However, even the most fervent promoters of central stimulants do not claim that ADHD is a life-threatening condition or that it shortens the life expectancy of the affected children. Therefore, these promoters, like the Swedish National Board of Health, are forced to play down the severity of the side effects and assert that they are relatively unimportant. Doing so, they also must pretend not to know of all the studies that show the opposite.

One thing they cannot deny is the result of the latest follow-up study of children treated with central stimulants, the MTA study from 2007. As stated above, the study showed that after thirty-six months of treatment, there were no positive effects whatsoever, contrary to what the promoters had expected, and that central stimulants are connected with a more aggressive and antisocial behavior and increased risks of future criminality and drug abuse.

As already stated, the findings of the MTA study were confirmed in 2010 by the Australian long-term study in which 131 children who had been treated with central stimulants were followed. The study showed that central stimulants had nil results on academic performance and didn't improve behavior. However, the blood pressure was significantly increased, and children who had been treated with central stimulants were ten times more likely to be identified as underachieving compared to their peers.

In spite of these results, the prescription of central stimulants in Sweden continues to escalate, increasing the profits of the pharmaceutical companies and harming an increasing number of treated children.

At the same time, these drugs are becoming more and more popular among parents and teachers because in many cases they seem to have such miraculous short effects on children's behavior, making them more compliant and more willing and able to do as they are told and facilitating the intercourse with other children. Teachers may find that these drugs limit the turmoil in the classroom, making it possible for them to teach in a meaningful way. And they are usually pleased that the children do as they are told.

Teachers may certainly notice that children become less spontaneous and curious and want to withdraw from other children and stop playing with them. One teacher expressed her experience of these children in the following way: "The wheel is rotating but you notice that the hamster is dead." Even if teachers notice such negative effects of central stimulants, they often regard them as a price to be paid in order to be able to teach.

However, in many cases, central stimulants will worsen the symptoms of the children, making them more aggressive, depressed or even suicidal, or causing severe obsessive symptoms, and in such cases, the medication often will be discontinued. And as the long-term studies have demonstrated, there are no positive effects on behavior after three years.

Another way to view ADHD: an impaired maturing of the brain

The idea that ADHD is a genetically determined biological disorder justifies a ruthless abuse of an ever-increasing number of children. Sooner or later, such destructive ideas must become obsolete and be replaced by more constructive concepts, which will make it possible for children with challenges to get real help. I have seen it as my undertaking in this book to present my experiences of how rhythmic movement training can help challenged children and to explain what I understand to be its manner of action.

As we have seen, the brain of the newborn infant is undeveloped and needs stimulation from the senses in order to mature. The baby will get such stimulation by doing spontaneous rhythmic baby movements, which will stimulate the growth of nerve cells in the brain stem, the cerebellum, the basal ganglia, the limbic system, and the prefrontal cortex as well as the connection between these parts of the brain. Dopamine is an important substance in the brain, the function of which is to transmit the nerve impulses between the nerve cells, and it is especially important for the functioning of the basal ganglia, the limbic system, and the prefrontal cortex. When new nerve connections are created in these parts of the brain, the supply of dopamine will increase.

When the motor development of the infant is obstructed in some way, the maturation process of the brain may be impaired or delayed. In my opinion, the causes of such delayed maturation are principally environmental and not genetic: birth injury; microwaves; toxicity from heavy metals; food intolerance, especially to gluten; cultural and psychological factors, and so on. The delayed maturation of the brain may cause various challenges with motor function, attention, concentration, impulse control, learning, and so forth.

Rhythmic movement training and reflex integration in ADHD

Rhythmic movement training is based on the rhythmic movements of infants and stimulates the development of new nerve connections, for instance in the basal ganglia and the prefrontal cortex. Thereby the motor function of the child and the ability to sit still will improve, as will executive functions. The development of the nerve nets in the brain will also increase the supply of transmitter substances—above all, dopamine. The child needs to do rhythmic exercises and reflex integration exercises every day to stimulate the development of new nerve connections in the brain.

Evidently, such a development must take time. After a few months of training, and sometimes much sooner, an obvious improvement of the symptoms of ADHD is usually seen if the child has done the exercises at least five times a week. The brain needs to be regularly reminded of the new patterns of nerve firing. When the newly established nerve connections have been reinforced by the training, the new responses will be firmly established. And when the original nerve connections have been inhibited, the original symptoms will disappear. It usually takes a year, and longer in severe cases, before the child has gotten rid of most of his challenges.

A comparison between rhythmic movement training and medication

Central stimulants cause brain damage in monkeys. In children treated with amphetamines, drug levels on a milligram/kilogram basis can be as high as those reported to cause brain damage in animals. In monkeys, the dopamine-producing basal ganglia and prefrontal cortex are especially damaged by central stimulants. There is no reason to believe that children would not be affected in the same way.

Experience of rhythmic movement training indicates that the function of the basal ganglia and the prefrontal cortex will improve in the end and that the supply of dopamine is increased when the nerve nets are stimulated to develop.

Studies have shown that central stimulants obstruct learning and higher mental processes, including flexible problem solving. Studies have not been able to show that central stimulants improve academic performance.

Experience of rhythmic movement training shows that learning is facilitated, especially in reading and reading comprehension but also in mathematics. Both in the short and long run, academic performance will improve considerably.

Central stimulants promote stereotyped, obsessive-compulsive, overly focused behaviors that are often repetitive and meaningless, and they suppress normal spontaneous or self-regulated activity, including curiosity, socializing, and play.

Experience of rhythmic movement training is that children become more extroverted and that their contacts with other children and their social ability will improve. This is also the case in autistic children. A scientific study of a group of chronic schizophrenic patients showed that rhythmic movement training caused them to become more interested in people around them and more ready to take part in social activities.

At the NIMH conference in 1998, a unique long-term study was presented, which showed a significant correlation between stimulant treatment in childhood and later use of (e.g., regular smoking, cocaine dependence, and a lifetime use of cocaine and stimulants). Also, the recent MTA study warns about increased risk of future drug abuse in children treated with stimulants.

Rhythmic movement training does not lead to any increased risk of future drug abuse.

Case study: Kalle

Kalle was eleven when he started rhythmic movement training in August. He had taken amphetamines for five years and now took 25 mg daily.

Kalle was early to develop and learned to walk when he was ten and a half months. He was very defiant when he was two to three years, and when he was six, he had such severe challenges with hyperactivity, concentration, and temper tantrums that he was given medication with an amphetamine. The first two years at school, he managed to stay in an ordinary class with a personal assistant, but then he was moved to a special class with seven students and five adults.

During his first visit to me, it was apparent that he was a hyperactive boy with great problems sitting still. His endurance was poor, and he was extremely impulsive and easily disturbed. Nearly every day he had severe temper tantrums. He had great problems playing with other children, and team sports were especially challenging. He preferred to play on his own and was constantly busy playing Game Boy or similar games. Kalle had several retained primitive reflexes: the Moro reflex, the TLR, the STNR, and the spinal galant reflex. He had no problem doing the rhythmic exercises and seemed to enjoy them.

Kalle did not like his medication and wanted to stop taking it and go back to his ordinary class at school. He seemed to be very motivated for these goals.

He was instructed to do passive and active rhythmic exercises every day and exercises for the Moro and the TLR reflexes a few times a week. In addition, his parents were encouraged to reduce his medication from five to four tablets a day. At his next visit, his father reported that he had not been motivated to do the active exercises every day, but he had always enjoyed being rocked and had done the active exercises when he felt like it.

After a little more than half a year, he had become more motivated to do the rhythmic exercises himself and reported that he felt calmer and more harmonious than before. He had begun to seek body contact with his parents and wanted to sit on their laps, which he had never done before. His temper tantrums had diminished, and his parents observed that he had become more reasonable and that he could follow a line of argument in a way he had never been able to before. At school, his behavior change had also been noticed, and it had been decided that he should move to a big class after summer vacation. He was to be assisted by a resource person.

Before summer vacation, I had recommended that the parents reduce his medication by another tablet. However, at his next visit after the summer, his parents reported that the school had protested strongly when they heard he was going to reduce medication and had even threatened not to move him to a normal class. Therefore, the parents had not reduced his dose. After that, it was more difficult to motivate Kalle to do the exercises, and only after persistently reminding him, he had done them about every third day. It was now decided that his medication should be reduced by one tablet immediately and then again after his next visit to me if he had done the exercises at least five times a week.

At his next visit in November, Kalle had reduced his dose by one tablet and his behavior was completely changed. Previously it had been difficult to interrupt his playing Game Boy in the waiting room. Now he romped and wrestled on the floor with his little brother whenever he had an opportunity. His parents reported that he had now taken responsibility for the rhythmic exercises and had done them every day on his own. His parents only needed to remind him. Assisted by his father, he regularly took part in the reflex integration exercises.

When he had done rhythmic exercises for one and a half years, his medication was reduced to two tablets a day. No one noticed any difference. He succeeded beyond all expectation at school. He continued to do the rhythmic exercises on his own and the isometric exercises with his father, integrating his TLR, spinal galant, and Moro reflexes. He had no more problems with endurance, concentration, attention, or hyperactivity. His temper was even, and he had no temper tantrums. Before summer vacation, he got nothing but praise from his teachers. He now reduced his medication with another tablet. In December, after doing the exercises for two years and three months, he took the last tablet. In February, two and a half years after his first visit to me, he reported that he felt good and school went well in spite of the fact that he had not done the exercises for several months.

After another year, Kalle read in the newspaper that persons who had been diagnosed with ADHD needed a psychiatric evaluation before they could get a driver's license. He then made an appointment with the child psychiatric clinic that had diagnosed him with ADHD and prescribed medication. He asked them to cancel his diagnosis since he no longer had any symptoms. After doing all the required tests, they had to cancel the diagnosis.

CHAPTER 4

Environmental Causes of Attention and Learning Problems

Genetic or hereditary?

When I first started to work with children with attention problems and other challenges, I noticed that their parents usually had had similar difficulties as children and in many cases had the same primitive reflexes active as their children. This confirmed that there are hereditary causes of ADHD. However, the theory that ADHD and similar conditions are exclusively hereditary and brought about by genes cannot explain the rapid increase of such conditions during the last two decades. An epidemic of a hereditary genetic condition can never occur. This does not mean that genes play no part in the increasing incidence of ADHD or autism. In 2010, a study was published that showed that autism may be caused by genes, the majority of which were not inherited from the parents but originated in the sperm or eggs.

If the damaged genes are not inherited from the parents, the environment must have caused the damage. These findings underscore the importance of environmental factors in the rapid increase of conditions such as ADHD and autism.

Children are affected by our harmful environment

Never before have we lived in an environment as harmful to our physical and psychological health as we are today. Environmental pollution; mercury and other heavy metals; vaccines; unhealthy food with abundant sugar and other sweeteners, MSG, and other food additives; and last but not least, increasingly intensive irradiation from mobile phones, mobile masts, cordless phones, and other cordless technology are only some of the growing threats to our health.

Fetuses, children, and adolescents are especially sensitive to this environmental disaster. Their immune systems and central nervous system are more sensitive than in adults, and the number of children who are born with malformations or who develop allergies, depression, attention deficit disorders, or autism is ever increasing.

Statistics showed an increase of 15 percent of malformation in Swedish fetuses, and between 1999 and 2004, children had an increase of 16 percent in chromosome aberrations.[29]

Young Swedish people of today are the first generation whose sense of well-being has deteriorated compared to that of their parents at a corresponding age:

- There has been a fourfold increase of suicide attempts in young people since 1980.
- There has been a threefold increase in young people who feel apprehension and suffer from anxiety.
- There has been an almost threefold increase in young people who suffer from insomnia, and more young people feel constantly tired.[30]

Mercury

Mercury is one of the most toxic substances that exists and is especially detrimental for the brain and the nervous system. Human health is affected by mercury from three principal sources, namely amalgam, the atmosphere, and vaccinations.

Since the nineteenth century, Amalgam has been an important source of mercury and is now recognized as a major cause of health problems in children and adults. According to an official Swedish report on dental material amalgam, it results in "risks of disorders of the fetal development, especially the development of the nervous system."

Another source of mercury is the burning of coal and garbage, which has resulted in a threefold increase of mercury in the atmosphere within the last hundred years. The US Environmental Protection Agency estimates that 15 percent of American infants have been exposed to risky levels of mercury during pregnancy.

In hundreds of studies, thimerosal, a mercury preservative in vaccines, has been linked to wide range of neurological disorders, and it has also been identified as one of the principal causes of autism.

Aspartame

Dozens of scientific studies within the last twenty years have demonstrated that the sweetener aspartame, most common in diet drinks, is one of the most harmful food additives. One of its metabolites (aspartic acid) destroys brain cells, especially in children. In addition, cancer and fetal damage are caused by another of its metabolites, methanol. Heavy consumers of diet drinks, including pregnant women, may consume thirty-two times the US government safety standard of methanol.

Phenylalanine is a metabolite that decreases the content of serotonin in the brain and causes depression, which may be one of the causes of the increase of depression in young people, especially girls who consume diet drinks in large amounts.[31]

Colorings and food additives

Food additives like MSG (monosodium glutamate) and colorings have long been known to produce symptoms of hyperactivity in children. In 2000, a scientific study was made that showed that the amount of colorings and food additives that British children on average consume every day causes allergies and behavior disorders such as hyperactivity, attention problems, and fits of emotion. A new study in 2007 confirmed these findings.[32]

Electroagnetic irradiation and electric smog

The most rapid and dramatic change of our environment is the staggering increase of electromagnetic irradiation from mobile phones, mobile masts, cordless phones, and other cordless technology, such as cordless broadband and so forth. There is by now a great amount of scientific studies showing that this is the most serious environmental disaster.

Children and young people are more sensitive to mobile irradiation than adults are. Research in Sweden has shown that there is an increased risk of brain tumors among mobile phone users between the ages of twenty and twenty-nine years and that the risk is greatest for those who started to use mobile phones before the age of twenty. For young users, the increase of the risk is 370 percent.[33]

A group of researchers at Lund University in Sweden has done studies in which young rats were exposed to irradiation from a switched on mobile phone for two hours. They could show that such exposure causes permanent brain damage, especially in the neocortex, the hippocampus, and the basal ganglia. Damage of these areas in children and young people is likely to cause problems with attention, concentration, hyperactivity, and learning.[34]

These misgivings have turned out to be true according to a Danish study in which thirteen thousand children born at the end of the nineties were followed up until the age of seven. Children whose mothers had used a mobile phone during pregnancy had a 54 percent increased risk of behavioral disorders, a 35 percent increased risk of hyperactivity, and a 25 percent increased risk of emotional problems. It was sufficient for the mothers to use the mobile phone twice or three times a week for the risk to increase. In addition, if the child had been exposed to a mobile phone after birth, the risk of behavior disorders increased 80 percent.[35]

As stated, electromagnetic irradiation may cause attention and learning problems by direct damaging the brains of fetuses and small children. Another mechanism also affects learning and attention. Common symptoms in ADHD are concentration difficulties, poor working memory, irritability and temper tantrums, poor endurance, sensitivity to sound and visual impression, and a tendency to be easily disturbed. The same symptoms are typical also in electro sensitivity. In both ADHD and electro sensitivity, these symptoms are caused by an increased stress level due to the fact that the irradiation triggers the stress reflexes, the fear paralysis and the Moro reflexes.

There has been a noticeable increase in behavioral and learning problems in children during the last fifteen years, coinciding with escalating exposure of children to electromagnetic irradiation from mobile masts, cordless technology, and the use of mobile phones. A primary school teacher who attended one of my courses in Helsinki asked me about the rapid increase in behavioral problems she had observed in her students. Previously, the increase was rather slow, but during the last years, the curve pointed straight up, she said. When I pointed to mobile phones and cordless technology as the most probable cause, she fully agreed with me and told me about her class of twenty seven-year-old students who all seemed to have enormous behavioral problems and learning difficulties. Eighteen of these students had their own mobile phones, and all of them insisted on having them on during the lectures in order to be able to have contact with their mothers all the time.

The Freiburg Appeal

An increasing number of doctors have noticed an alarming upsurge of severe diseases and symptoms during the last decade. In 2002, a group of German doctors published the Freiburg Appeal, which has been signed by more than one thousand German doctors. These doctors had "noticed a dramatic increase of severe and chronic diseases" during the last few years, including, but not limited to learning and concentration problems and behavioral disorders in children (e.g. hyperactivity) besides cancer, leukemia and brain tumors.

The appeal continues: "In addition, we more often observe different disorders, which are incorrectly perceived as psychosomatic, e.g.: Headache, migraine, chronic fatigue syndrome, insomnia, tinnitus, nerve pain and pain of soft tissue."

The appeal also calls attention to the fact that an accumulation of such cases can be observed in areas with strong irradiation and that the symptoms often improve or disappear when the irradiation is decreased or eliminated.

How does the medical profession respond?

When children are affected by our harmful environment and show various symptoms, most doctors do not respond like the German doctors who signed the Freiburg Appeal. Instead, they respond by giving them psychiatric diagnoses and medication. A prominent Swedish psychiatrist argues that psychiatric disease is far more common among children than has previously been recognized. According to this expert, epidemiological studies have shown that the rate of psychiatric disease among children amounts to 20 to 25 percent, with 6 percent considered to suffer from depression and 7 percent from ADHD. According to other experts, one child out of ten suffers from ADHD.[36]

As previously stated, diagnoses are social constructions. The medical profession creates them by combining various symptoms and declaring these combinations to constitute diseases. Children who suffer from attention deficit disorder, impulsivity, and hyperactivity are considered to suffer from ADHD (attention deficit hyperactivity disorder). Although there is no consensus within the medical profession about the cause(s) of ADHD, the recommended treatment of ADHD with central stimulants like Concerta and Ritalin is not questioned.

Children and young people who are irritable, passive, and sad, who suffer from lack of energy, tiredness, and insomnia, are diagnosed with depression. Doctors normally don't question the treatment with antidepressants in spite of the fact that scientific studies have demonstrated that depression in many cases is a symptom of electromagnetic radiation or gluten sensitivity.

How do the media and politicians respond?

In the media, at least in Sweden, there is seldom any serious discussion about the causes of the deteriorating health of children and young people. Most commonly, the media emphasizes the increased stress in society as a cause. Scientific studies that demonstrate how children are harmed by radiation, vaccinations, medicines, or food additives are covered up or hushed up, while other studies that show no harmful effects are uncritically blown up. Reports about effective and safe methods to help children with problems are conspicuous by their absence.

In Sweden, the government treats these problems by increasing the grants to child psychiatry so more children can be diagnosed and treated with drugs.

That the health of children is continually deteriorating is as big a threat to our future as global warming is. That there is no serious discussion about this in the media and the fact that no responsible authorities are prepared to treat these problems seriously is bizarre. It is as bizarre as it would be if the media did not discuss the causes of global warming and the politicians just ignored the issue.

Who needs diagnoses?

As previously stated, diagnoses are social constructions. To uphold such constructions, experts who are authorized by society to diagnose are necessary. These experts are normally doctors, but the rapid deterioration of the health of children and especially the increase of children with behavioral disorders has made it necessary, at least in Sweden, to train a body of psychologists and social workers to diagnose children who react to their harmful environment.

As social constructions, the diagnoses meet different needs. Such needs may be economical, political, professional, or psychological. The pharmaceutical companies need children to be diagnosed for economical reasons in order to be able to sell drugs as central stimulants and antidepressants. Scientists, doctors, psychologists, and social workers need the diagnoses to justify themselves in their respective professions, to get the politicians to make grants for their work and research, and to get posts and to make careers.

Politicians need the diagnoses to cover up the fact that the deterioration of children's health is largely caused by environmental factors that they are responsible for. Parents and teachers need the diagnoses to get support and resources and escape feeling like failures.

Children, on the other hand, do not need diagnoses. Children who have problems with attention and hyperactivity do not need to be treated as suffering from a disease that, according to experts, "seriously affects their health and development, undermining their basis for satisfactory lives as adults." They do not need to hear that they run the risk of academic and professional failure and of serious social and psychiatric problems as adults.[37]

Children with challenges need help. They need real help to feel and function well. They do not need to be stigmatized by diagnoses and poisoned by drugs that subdue their symptoms and are harmful to their systems.

A biologically and ecologically sustainable environment

Primarily, children need biologically, ecologically sustainable, and healthy environments. It should be the responsibility of society to provide for this, but this is not done. And when politicians, health, authorities, and the medical profession let children down and sacrifice their health, the only way out is knowledge and individual initiative from parents and teachers and all people around children to improve the environment of children and improve their health in the end as well.

Some suggestions to begin with:

- Clean up the environments of fetuses, infants, and small children from unnecessary electromagnetic irradiations like cordless technology and wireless broadband. Never use cordless DECT telephones in their vicinity and mobile telephones closer than two meters from them. Pregnant mothers should never use mobile phones, cordless phones, or cordless technology.
- Restrict the use of mobile phones and cordless technology by children and young people and never allow them to use cordless DECT telephones.
- Eliminate sugar, sweeteners, MSG, colorings, and other food additives from the food of children. Make sure their diets meet their need for minerals and vitamins. Give them supplements of vitamins, minerals, and omega-3 fatty acids if this is indicated.
- Search for alternatives to central stimulants like Concerta and Ritalin and to antidepressant drugs that doctors prescribe.
- Study the risks of vaccinations and think about if and when you want your child to be vaccinated.

Chapter 5

The Brain Stem and Rhythmic Movements

The neural chassis and its three guiding operators

Paul MacLean, who has studied the development of the brain in amphibians, reptiles, and mammals, has described how the different levels of the brain cooperate. The nerve chassis in which he includes the spinal cord and the brain stem "provides most of the neural machinery required for self-preservation and the preservation of the species. By itself, the neural chassis might be likened to a vehicle without a driver. Significantly, in the more advanced vertebrate, the evolutionary process has provided the neural chassis not with a single guiding operator but rather a combination of three, each markedly different in its evolutionary age and development, and each radically different in structure, chemistry, and organization." The three operators that guide the neural chassis are the reptilian brain, or the basal ganglia; the mammalian brain, or the limbic system; and the neocortex.[38]

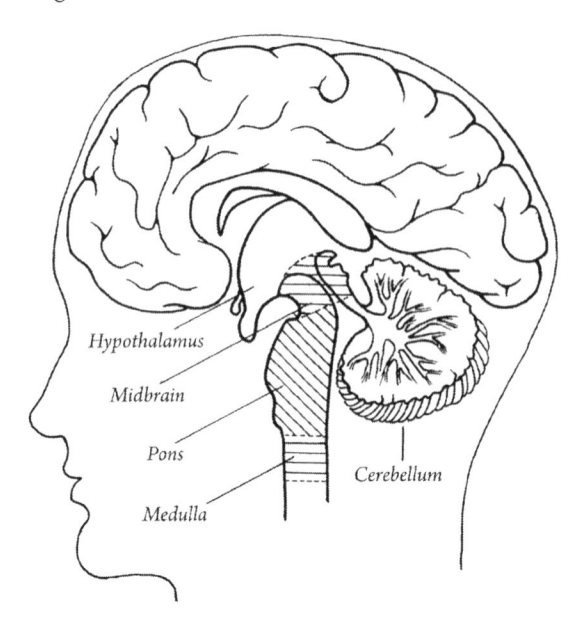

Fig. 3 The brain stem and the cerebellum

The development of the motor abilities and the senses of the fetus

In the fetus, the brain stem develops gradually and must function adequately for the newborn baby to survive. Like the brain of the fish, the brain stem of man receives signals from the senses: the vestibular, kinesthetic or proprioceptive, visual and tactile senses. The brain stem then responds by relaying signals to the motor organs. In addition, respiration, heart activity, and other life-sustaining processes are directed from the brain stem.

In the fetus, the senses mature gradually, and at birth they are still undeveloped and not adapted to life outside the womb. The newborn infant is completely dependent on his parents. The proprioceptive sense enables the fetus to keep track of its different parts of the body. The fetus learns to do well-adjusted movements in the water environment of the womb and is able to move freely and both suck its thumb and play with the umbilical cord. In the new gravitational circumstances after delivery the baby looses these abilities and moves like a fish on dry land with slow movements mainly with the head, trunk and arms. The newborn baby is still a good swimmer.

The vestibular sense develops early in the fetus and enables both the fetus and the newborn baby to keep her balance in water. On dry land, however, the baby has not developed any sense of balance.

In the very beginning, the motor abilities of the fetus are controlled from the spinal cord, involving simple reflex movements. When the brain stem gradually starts to function, the more complex reflex patterns of the primitive reflexes are developed, which will help the fetus and the infant to develop its motor abilities further. However, no voluntary motor abilities have yet been developed.

The newborn infant is faced with the task of reprogramming both its vestibular and proprioceptive sense and to learn how to control its motor organs and move freely on land. In order for the infant to go about this reprogramming, it must make spontaneous, rhythmic infant movements according to its inborn program.

The brain stem and muscle tone

All senses except the sense of smell send sensory signals to specific areas of the brain stem, called nerve nuclei. These nuclei process the information from the senses, integrate it with information from other senses, and send it on to higher levels of the brain. The vestibular nuclei that receive signals not only from the vestibular sense but also from other senses, especially the tactile and the kinesthetic ones, are particularly important for muscle tone. If there is insufficient stimulation of the brain stem from the tactile, kinesthetic, and vestibular senses, the result will be low muscle tone of the extensor muscles of the body. Therefore, it is of crucial importance that the baby is touched, hugged, rocked, and allowed to move around freely. Such stimulation sends signals from the sense organs to the vestibular nuclei. Lacking such stimulation, the baby may develop low muscle tone and have difficulties lifting his head and chest and moving around, further reducing the stimulation from the vestibular, tactile, and kinesthetic senses, causing a vicious circle.

Due to muscular weakness, such a child may have problems keeping his head in an upright position and may easily shrink up. Often the joints and especially the spine of such a child are overly flexible. Due to the hunched-up posture, the breathing is shallow. The child may be unwilling to move around and may prefer to sit still.

Many such children will develop attention problems due to malfunction of the cortex of the brain and be diagnosed with ADD when they grow older. Overly flexible spine and joints make it even more difficult for them to keep up proper posture and breathing, causing insufficient arousal of the neocortex by way of the reticular activation system.

The rhythmic movement training will gradually normalize muscle tone and decrease the over flexibility of the joints.

The brain stem and the reticular activating system (RAS)

The central part of the brain stem mainly consists of a dense nerve net, the reticular activation system (RAS). This system of nerve cells receives signals from the visual, auditory, and vestibular senses; from the sense organs of muscles, joints, and inner organs (the proprioceptive sense); and from the tactile sense, transmitting the information to the cortex. The effect of such signals is to arouse the cortex, and they are necessary for maintaining attention and alertness. Without such arousing influence on the cortex, we are not able to maintain awareness of external events

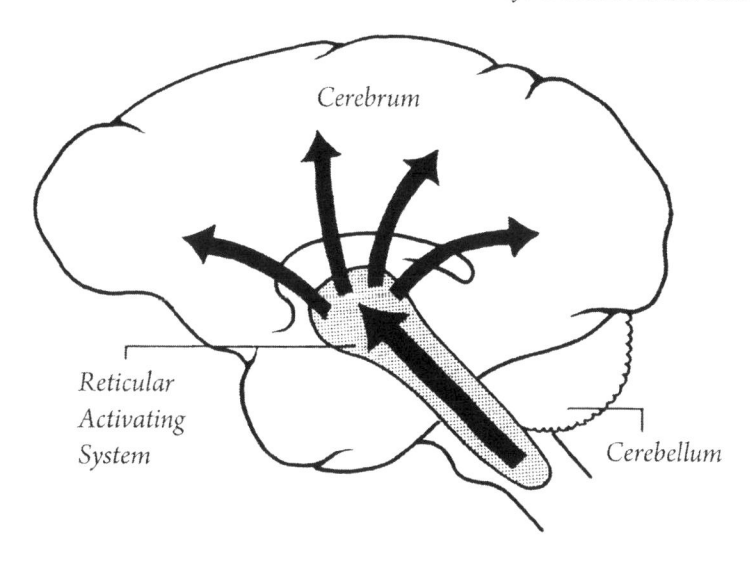

Fig. 4 The reticular activation system (RAS)

Experience from rhythmic movement training indicates that a constant lack of signals from the vestibular, proprioceptive, and tactile senses causes deficient alertness and attention, which will rapidly be remedied by rhythmic stimulation. Children with low muscle tone and hunched posture may develop attention deficit disorder (ADD) without hyperactivity. In extreme cases, children with very low muscle tone, who are unable to move around, may get so little arousal that they are unable to maintain awareness of external stimuli and instead habitually daydream or even hallucinate.

The primitive reflexes

The motor activity of the fetus depends on primitive reflexes. These are automatic stereotyped movements controlled from the brain stem. These reflexes are developed during different stages of pregnancy and must mature and finally be inhibited by the basal ganglia and integrated into the whole movement pattern of the baby. By making rhythmic baby movements, the baby inhibits and integrates these reflexes one after the other.

The primitive reflexes are triggered by sensory stimulation from the senses. One of the earliest reflexes to develop is the tonic labyrinthine reflex (TLR). When the head of the fetus is bent forward, the spine and the limbs are also bent. After birth, when the baby extends his head backward, the TLR backward is developed and the baby stretches his body.[39]

When the fetus turns its head from side to side, proprioceptive receptor organs of the neck will trigger the asymmetric tonic neck reflex (ATNR). This happens after eighteen to twenty weeks of pregnancy, causing the fetus to stretch his arm and leg to the side to which the head is turned. The mother then notices that the baby starts kicking.

The development of the automatic gait reflex into the mature adult walk may illustrate the role that the maturing brain plays in the developing of our motor abilities. The fetus starts to make walking movements very early, even before the brain stem and the spinal cord have been linked up. Such gait movements are controlled from the spinal cord.[40] Later during pregnancy, the primitive automatic gait reflex is developed in about the thirty-seventh week. This reflex is controlled from the level of the midbrain (upper part of the brain stem) and is active when the baby is born. It can be triggered by holding him under the arms and lifting him into an upright position, leaning him slightly forward, and letting the soles of his feet touch a flat surface. Then the baby starts making automatic gait movements.[41]

When the baby is three or four months old, the automatic gait reflex is inhibited and can no longer be triggered. However, it is not until the baby has learned to rise up into a standing position and master gravity forces that he will be able to transform the automatic gait reflex into the mature adult walk reflex, a postural reflex controlled from the basal ganglia.

The importance of the maturing of the primitive reflexes

If the baby is unable to inhibit his primitive reflexes in time, they will delay his motor development and consequently obstruct the maturing of his brain.

An infant in whom the primitive reflexes have not properly developed and matured when it is born will have more difficulties inhibiting them in due time than a baby whose reflexes are fully matured. This is especially the case in premature babies and babies delivered by cesarean section.

In premature babies, several primitive reflexes may not have developed when they are born. In an incubator, the baby does not get stimulation from the tactile, vestibular, and proprioceptive senses as in the womb. This will delay the maturing of primitive reflexes. If, on the other hand, the mother carries the premature baby around on her chest, it will get similar stimulation as in the uterus, which will promote the maturing and integration of primitive reflexes and improve muscle tone and help the linking up of the different levels of the brain.

Many babies nowadays are delivered by caesarean section and therefore miss the normal delivery process. It is during the birth process that many of the baby's primitive reflexes are triggered, which is important for the maturing of the reflexes. Premature babies and babies delivered by caesarean run a greater risk of retaining their primitive reflexes into adulthood and acquiring problems with mobility, attention and concentration, and learning due to incomplete maturing of the brain.

Rhythmic movements for stimulating the brain stem

The fetus gets sensory stimulation from the mother's respiration, heartbeat, walking, running, and so forth. Such passive stimulation affects the tactile, proprioceptive, and vestibular senses of the fetus and stimulates the growth and maturing of the nerve cells of the nerve chassis. All such stimuli also promote the maturing of other parts of the brain.

Also, when primitive reflexes are triggered in the fetus, the motor responses cause stimulation of the nerve chassis. Another source of sensory stimulation is all movements the fetus can do by itself, such as turning its head from side to side, sucking its thumb, playing with its umbilical cord, and so on. Playing with the umbilical cord will cause proprioceptive stimulation, which has a calming effect on the fetus.

Passive rhythmic movements are especially useful to stimulate the nerve chassis in infants and in children with brain injuries who are still neurologically on the level of infants. By passively rocking the child in different ways, the nerve chassis will be stimulated to improve muscle tone, maturing primitive reflexes and stimulating spontaneous movements. When an infant is slow to develop and does not easily move on from one stage of development to another, such stimulation can be used to quicken development (e.g., with children who are not able to lift their heads or do not start crawling on hands and knees).

In a child with severe brain damage who is stuck with his head turned to one side, passive rhythmic stimulation may cause him to turn his head from side to side in a reflex-like manner.

Case study: Olle

When I followed the work of Kerstin Linde, I met Olle, who was five years old when he started to do rhythmic movement training. He had stopped developing during the previous year. I followed his progress for half a year, during which he made great progress. When he was born, he was quite floppy and limp and had not even enough strength to cry. He developed slowly. When he was eight months old, he was still lying on the floor and had only learned to raise his head a little. The parents were not able to make eye contact with him. He was tired and sluggish. When he was eleven months, he learned to sit, and when he was eighteen months, he began to rise and even learned to stand for some time. But as he grew and got heavier, he lost his ability to stand.

The medical establishment gave the parents no hope at all that Olle would improve. They got the impression that the staff had the attitude that there was nothing that could be done to help Olle, and they shrugged their shoulders at the whole thing. Olle's parents felt that the physiotherapists did not believe that they could help him, and they only saw him once a month, as was their duty.

Just before he visited Kerstin Linde for the first time, he used to sit quite still most of the time and he hardly made any sounds. His back was bent and shrunken, and he had no strength to hold his head in an upright position. His legs were quite limp. He did not meet his parents' eyes and hallucinated a lot. He was not interested in his surroundings except for those times when his parents played music or sang for him, which always animated him.

Olle was given the basic movements to begin with. He was very positive to the movements and was diligent about doing them: kneeling on all fours, rocking backward and forward, and crawling on all fours in the apartment. He started to pull down things and showed an interest in his surroundings by putting things into his mouth. He suddenly wanted to eat by himself, something he had never done before. He started to babble and make different sounds, and he started to pay attention when he was called by name.

His vision improved, and he began to look fixedly at things around him in a totally new way. Previously he had mostly shifted his gaze and only absentmindedly watched TV. Once he started to watch TV more than before, he discovered his hands, which he started to examine, and he became very interested in looking at himself in the mirror, to which he had never paid any attention previously. For the first time, he noticed the family dog and started to play with it. When the family received visitors, he showed interest and crawled up to look at them. His hallucinations decreased. He began to react when he hurt himself, something he had never done before.

After a few months of training, his posture had completely changed. He managed to hold his head in an upright position. His back was not bent as it used to be, and his legs were not so floppy.

Olle visited once a month, and one or two days after every visit, he used to become ill. After the first visit, he got a fever and sore throat and expectorated a lot of phlegm. After a couple of months of training, there was a terrible rattle in his chest; his parents became very worried. However, suddenly he was able to cough for the first time in his life so he could cough up the phlegm.

After a couple of months of the training, he started to have emotional reactions. He began to get a will of his own. He learned to protest if he did not want some food, and then he could even shake his head, which was something entirely new. And if he wanted something, he did not give up—he become extremely stubborn. Previously he had been indifferent and the parents had been able to stuff him with anything. He would just open his mouth and swallow.

Five months after Olle started his training, his physiotherapist noticed his progress and that his body had become much steadier. By this time, he had not visited her for three months. Then she said that she wanted to see Olle more frequently. His parents did not dare tell her that he was going to an alternative place for treatment. They chose to stop seeing Kerstin Linde and see the physiotherapist instead.

Chapter 6

The Cerebellum and Rhythmic Movements

The cerebellum is situated like a bulge from the brain stem. The cerebellum receives signals from the kinesthetic sense and from tactile sense receptors that transmit information about touch and pressure. Between the cerebellum and the motor cortex, there are important nerve connections that enable the cerebellum to play a fundamental role in coordinating our movements.

Together with the vestibular nuclei, the cerebellum correlates postural and kinesthetic information about the position of the body in the gravitational field. Its function is to make movement smooth, easy, and coordinated, correcting deviation between executed and planned movement.

The cerebellum plays an important role for the rhythms of the brain and the body. The rhythmic elements of the spontaneous infant movements seem to have a special importance to stimulate the growth and maturing of the nerve nets of the cerebellum and the brain as a whole. During the second half of the first year of life, there is rapid growth of the cerebellum, coinciding with fast movement development of the baby.

Harald Blomberg, MD

Dysfunction of the cerebellum in attention and learning disorder

Some children may have great difficulties making simple rhythmical movements in a coordinated way. This inability is usually not apparent when the child moves in an upright position and is therefore mostly overlooked. These children may be unable to make such simple rhythmic movements in an active way. For example, actively rolling the bottom from side to side (movement number 9) or actively sliding on ones back (movement number 7) may be hard to learn for some children. Such an inability may reflect a dysfunction of the cerebellum, which is likely to affect the function of many other parts of the brain, especially the cortex and many areas of achievement. It may affect attention, planning, judgment, control of impulses, and abstract thinking. It may affect eye movements, reading comprehension, speed of information processing, working memory, learning, and speech development.

The great impact of the cerebellum on all these functions can be explained by the strong connections between the cerebellum and those areas of the neocortex essential for such faculties. Such connections exist between the cerebellum and the prefrontal cortex,[42] responsible for attention, planning, judgment, and control of impulses; between the cerebellum and the speech areas of Wernicke and Broca;[43] and between the cerebellum and the area in the frontal lobes, responsible for eye movements. When these areas get insufficient stimulation from the cerebellum, their nerve nets do not develop properly, which explains their poor function.

Rhythmic movement training and the cerebellum

A strong impact on attention, concentration, control of impulses, abstract thinking, judgment, and learning is usually seen in rhythmic movement training. This can be explained by various factors such as improved arousal of the cortex from the brain stem or increased stimulation of different areas of the cortex by the cerebellum.

Children who have difficulty doing rhythmic movements in a smooth, rhythmic way due to a dysfunction of the cerebellum do not benefit as rapidly from the movements as children with no such difficulties. It is therefore most important to teach children with such challenges to make the movements in a rhythmic way. Some of them learn quite quickly, within a month or so, but others may have to practice daily for more than a year before they can make the movements smoothly, rhythmically, and effortlessly. Even then, they tend to lose the rhythm when they are tired.

Rhythmic movements that are made actively are most important for remedying any dysfunction of the cerebellum. In addition, these movements have other effects such as integrating primitive reflexes and developing postural lifelong reflexes, especially in small children. These movements also promote the linking up of different parts of the brain—for instance, by stimulating the growth of nerve nets that are essential for the arousal of the cortex and the stimulation of different areas of the cortex by the cerebellum. All these effects are most important for resolving attention and learning difficulties.

It takes time to rebuild the brain in this way, and therefore the brain needs continuous and daily stimulation from active rhythmic movements for a long time, most commonly a year or more before learning and attention problems can be completely resolved.

In children who are not able to do rhythmic infant movements due to some motor handicap (e.g., cerebral palsy or severe weakness of the muscles, as in Olle's case), the nerve nets of the cerebellum will not get sufficient stimulation. Consequently, the prefrontal cortex or the speech areas of the cortex will not get sufficient stimulation. In such cases, speech may not develop and there may be great problems with attention. When such children start to do rhythmic movement exercises, there may be rapid development due to the stimulation of the cerebellum.

Dysfunction of the cerebellum due to inflammation

Dysfunction of the cerebellum may not only be caused by insufficient stimulation due to deficient motor development. The cerebellum may also be damaged by inflammation, causing symptoms such as delayed speech development, problems with articulation, attention difficulties, and so forth. Such inflammation is normally seen in autism but is becoming more common in children who are not on the autistic spectrum. In most cases, the inflammation is caused by gluten sensitivity, which is becoming more and more common. Scientists estimate that between 10 and 30 percent of children may suffer from gluten sensitivity, which could explain the rising frequency of late speech development and attention problems in children nowadays. The cerebellum is damaged by antibodies against gliadin (a component of gluten), which travel in the bloodstream to the brain, where they cause an autoimmune reaction, especially in Purkinje cells of the cerebellum.

In my experience, children with poor articulation and late speech development are usually gluten sensitive and need to be on a gluten- free diet in order to limit the inflammation of the cerebellum. Otherwise, the rhythmic exercises may not be very helpful. With a combination of rhythmic movement training and diet, children with late speech development usually develop normal speech within a few months.

Exact movements

In order for the rhythmic exercises to be as effective as possible, they must be done in a rhythmic, coordinated, and smooth way. In addition, they should also be done symmetrically. Unsymmetrical movements usually are signs of active primitive reflexes. Subjectively we experience our movements as symmetrical even if we only move one-half of the body (e.g., only rotate the head to one side and not to the other). Often you only need to call the client's attention to the asymmetric movement pattern for her to correct it.

When we do rhythmic exercises, there must be no accessory movements of the mouth, feet, hands, shoulders, neck, or hand. When the client's attention is called to such accessory movements either by pointing them out or by putting a hand on the body parts that should not move, he will quickly learn not to involve muscle groups that should not be active.

Movements that are made in a rhythmic, smooth, symmetrical, and coordinated way can be considered exact movements. Some individuals can make such movements spontaneously, while others may have to work hard to achieve such movements. The movement exercises must therefore be adapted to each individual so he can learn to do them more and more exact.

Infants will learn how to make the rhythmic baby movements exactly if they are able to and given the opportunity to move freely on the floor. When they develop a new movement, it first looks tentative, but gradually it will become more and more exact. Some children, however, will not give themselves sufficient time to develop a movement properly before they hurry up to do another one. Others may have more or less severe damage that prevents the infant movements from developing exactly.

The more exact the exercises are made, the more information they give the brain, making it possible for the infant to adjust muscle tone in each moment and enable the back and the joints to work in the best possible position and cooperate optimally in the gravitational field. In many cases, the spine or the joints may have become fixed in incorrect positions. For instance, the thoracic spine may have been locked in a hump or the pelvis may be rotated, just to mention two examples. Such fixations must be corrected before the rhythmic exercises can be done exactly.

Case study: Eva

One of the most instructive cases I observed during the three years I followed Kerstin Linde's work was Eva. Her motor development was extraordinary. She would most probably have learned to walk if it were not for the well meaning but nevertheless sadistic treatment she got in the rehabilitation ward. Also, her speech development was remarkable for a girl about whom the doctors had prophesied she would never learn to talk. Her rapid development of speech and motor abilities intrigued me and inspired me to find a way to explain her progress.

Eva had just reached the age of three when she visited Kerstin Linde for the first time. She was a lean girl with a thin face and small cold feet. She could turn from her stomach to her back but not the other way around. She could not sit up by herself; she had to be supported by pillows. She could not eat without help. She could not speak a single word.

When Eva was one year old, it was established that she had cerebral palsy due to lack of oxygen at birth. After that, she got extensive help from the medical services. She had sessions with a speech therapist and a physiotherapist, and she had her own assistant. Her mother told me, "She had the same physiotherapy program all the time, and her assistant made movements with her several hours every day for two years, but nothing happened. Eva kept lying on the floor and could neither rise nor turn around." The doctors had said that Eva would never learn to speak, and it had been decided that she should take part in a project to learn sign language.

Eva's mother told about their first visit. "Kerstin said that Eva's brain believed that she had only one leg and that it had to be informed that she had two. And then she took her two legs, bending and stretching them as she said 'Right, left, right, left'. Then Eva was told to pull her right leg and then her left one. And she did it! I could hardly believe my eyes; she had never done anything like that before. I was asked to keep a diary about what happened with Eva. I thought that nothing much would happen. If nothing had happened in the previous two years, I thought it would certainly be enough to write only once a week. However so much happened that I had to write one page every day in the diary."

In the evening when they got home after Eva's first visit, she managed to sit up on the floor. First she lay on her stomach, then she rose on all fours, and then she backed up and sat.

Not long after she started the training program, she began to use her hands so she could eat and drink by herself. After a year, she was able to do a jigsaw puzzle, dress and undress her dolls, and switch the tape recorder on and off. A couple of months after she started the training, Eva began to speak, first occasional words and then two-word sentences, and after a year, she was saying up to six-word sentences. Soon after she began the training, her legs were quite relaxed and she was able to lie on her back and suck her toes. It did not take long before she could sit on the floor and occupy herself. Eight months after she had started the treatment, she had learned to rise to her feet, supporting herself on furniture. Her vision had improved, and she had to change her glasses.

This is how Eva's mother described her training program: "In the beginning, we had to work hard to get her to relax. We took her by the feet and rocked her in the longitudinal direction of her body, and then we rolled her legs. When she was relaxed, we could start to do active exercises. She used to kneel on all fours and rock to and fro, and then she used to lie on her stomach and make crawling movements with her legs. Soon she learned to use her arms to move on the floor, and after some practice, she learned to crawl on all fours."

After eight months of treatment, just after she had learned to rise to her feet, Eva was admitted to the child rehabilitation clinic. Her mother reported: "The opinion of the medical staff was that Eva would never learn to walk. They had to develop whatever could be developed, they said. So she was tied to a board and had to stand up supported by the board or had to sit in a wheelchair the whole day. She was not allowed to be on the floor. The physiotherapists said that it was no use working with her legs until she had an operation on them, but the orthopedists did not think that an operation was required and refused to operate. Eva did not want to be tied up and was angry and protested. When I came to take her home, they said that she was not at all cooperative, and even among the worst children they had treated. They advised me to consult a child psychologist for Eva and even gave me the address of one. When Eva got home, she had regressed totally. Her legs were quite stiff, and she could not even rise from a prone position, something she had learned the first time she visited Kerstin. Not until we visited with her the next time did she get the help she needed so her legs could relax again."

Her mother reported her development after coming home from the child rehabilitation clinic: "When Eva had learned to drive her wheelchair, she discovered how comfortable and fast it was. When she got home, she did not want to be on the floor. She became much stiffer, and the training became much more difficult. After she began sitting in her wheelchair she also got several urinary infections, something she had never had before. But at the same time, Eva is so happy to be able to make her way more easily at home."

Needless to say, Eva did not learn to walk.

Eva's case illustrates how movement ability and speech are linked and how speech will only develop when motor abilities improve and the cerebellum is stimulated. Because of cerebral palsy, Eva had not been able to move as an infant and her cerebellum had therefore not received stimulation to develop. When she started rhythmic movement training, the cerebellum started to develop and could stimulate the speech areas of the left hemisphere, enabling speech to develop.

CHAPTER 7

The Reptilian Brain, or the Basal Ganglia

The functions of the reptilian brain

According to Paul MacLean, the human brain consists of three layers that surround the brain stem, or nerve chassis, which roughly corresponds to the brain of the fish. As stated, by itself the neural chassis might be likened to a vehicle without a driver. In the more advanced vertebrates, the evolutionary process has provided the neural chassis not with a single guiding operator but rather a combination of three. The reptilian brain, next to the brain stem, corresponds to the new structures of the brain that reptiles developed. It is the first part of the triune brain to develop to function as a guiding operator of the nerve chassis. In reptiles, this part has no demonstrable importance for motor ability according to Paul MacLean. In reptiles, this part of the brain controls social interaction, which is characterized by rituals, routines, and strict hierarchical submission.

In humankind's rituals, routines, and hierarchical dominance are also controlled by the reptilian brain, or the basal ganglia, as this part of the brain is called in man. But these behaviors are not as prominent in man as they are in reptiles. As frequently happens during evolution, a new function has been developed by an old structure. The basal ganglia in mammals and humans have been modified to have the main task of controlling motor activity, working in close cooperation with the motor cortex. The basal ganglia can be seen as an intermediary between the brain stem and the neocortex in their function to develop automatic postural reflexes, which make the control of movement in the gravitational field possible. Unlike the primitive reflexes, the postural reflexes can be governed to some extent by the motor cortex, thus permitting voluntary movement to take place.

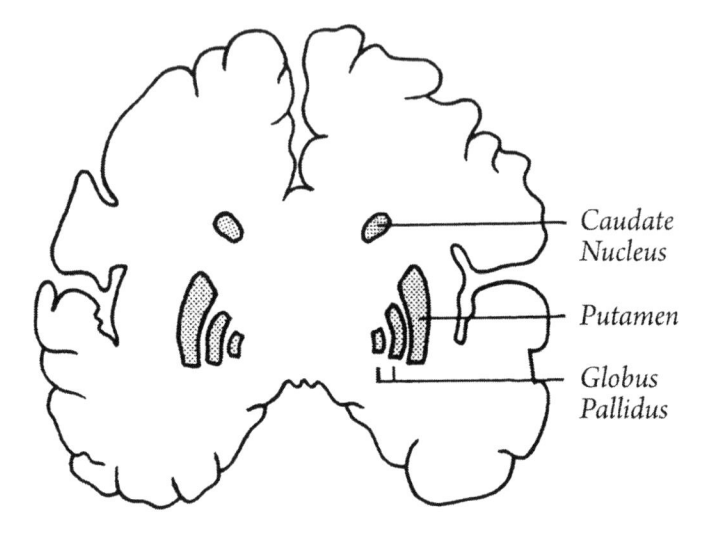

Fig. 5 Cross section of the brain showing the reptilian brain, or basal ganglia: nucleus caudatus, putamen, and globus pallidus

The postural reflexes

The motor activity of the newborn infant is controlled by primitive reflexes and is outside voluntary control. In order to control his motor organs voluntarily in the gravity field, the baby must develop new reflex patterns, the lifelong postural reflexes that are controlled from the basal ganglia.[44] When the baby makes his rhythmic baby movements according to an inborn program, the nerve nets of the basal ganglia are stimulated and the postural reflexes are developed.

The postural reflexes are necessary for our stability and ability to balance while kneeling on hands and knees or in sitting and standing positions. They enable us to move automatically by crawling, walking, or running and rising from sitting or lying positions, the so-called locomotion and rising reflexes.

The basal ganglia receive signals from the proprioceptive, tactile, vestibular, and visual senses and respond by sending signals downward to the brain stem and spinal cord (the nerve chassis). These signals inhibit or modify the primitive reflexes, transforming the stereotyped and sweeping movement patterns of the primitive reflexes into the more precise and well-balanced movement patterns of the postural reflexes. Therefore, the basis of the postural reflexes are the primitive reflexes that are inhibited and transformed by the basal ganglia and integrated into the movement pattern of the baby.

The inhibition and transformation of the primitive reflexes into lifelong postural reflexes is brought about by the spontaneous rhythmic movements that a baby makes before he learns to walk. At the same time, the spontaneous baby movements develop the postural reflexes. The main part of the inhibition of the primitive reflexes should be completed while the baby is still on the floor. Movements in upright positions, standing or walking, inhibit the primitive reflexes only to a limited extent.

The importance of the basal ganglia for automatic movements and the ability to sit still

Like the primitive reflexes, the postural reflexes are automatic and cannot be replaced by voluntary movements. When we move around, we cannot be aware of every single movement. We need to decide voluntarily when we want to start moving, where we want to go, and how fast we want to move.

When we are at rest, the basal ganglia are in full activity and have a strongly inhibiting effect on our movements. When we decide to start moving, the motor cortex sends signals to the basal ganglia to decrease this inhibiting effect. The quicker we move, the less inhibiting is the effect of the basal ganglia.

Small children who develop normally are usually overactive and move around more or less constantly, except when they are engaged in some very interesting activity. Like older children who are overactive, they cannot be still by command. This is because the nerve nets of the basal ganglia are undeveloped in children who are still learning to master their balance and stability.

Not only the automatic postural reflexes but also all other automatic movement patterns that have been learned are controlled by the basal ganglia. They have therefore been called the "secretaries of the brain." In children whose basal ganglia are undeveloped (e.g., small children or those with ADD/ADHD), the ability to learn to do things automatically is impaired. Children with ADHD must compensate for this problem by being conscious of what they are doing all the time, which may contribute to their poor endurance and make them tire quickly of what they are doing.

The basal ganglia and Parkinson's disease

The postural reflexes are developed on the basis of the primitive reflexes. Even when the postural reflexes have been established, the original primitive reflex patterns remain on the level of the brain stem, although normally they should not be active. These reflex patterns reappear when the inhibiting effect of the basal ganglia end for some reason or other. This is exactly what happens in Parkinson's disease.

In 1967, British neurologist Dr, Purdon Martin published a report on 130 cases of Parkinson's disease caused by an epidemic of encephalitis between 1919 and 1925. He was able to show that such late symptoms of the disease as inability to keep stability or balance, to rise and sit down and to walk normally, are due to dysfunction of the postural reflexes and a consequence of the extensive death of nerve cells in pallidum, one of the nuclei of the basal ganglia.[45]

Purdon Martin did not investigate what happens with the primitive reflexes in Parkinson's disease. However, my own experience of treating clients with Parkinson's shows that the primitive reflexes are activated early, before the postural reflexes have been affected. By doing movements for integration of primitive reflexes and training of postural reflexes, motor abilities improve and the longtime impairment of motor ability that usually happens in Parkinson will not take place.

It is not only in Parkinson's disease that the primitive reflexes may be reactivated later in life. Old age as well as whiplash injury and other traumas are also able to reactivate the reflexes.

Motor development of the baby

Primitive reflexes control the motor activity of the newborn. When the baby makes spontaneous rhythmic infant movements, the primitive reflexes are inhibited and the postural reflexes are developed. After birth, the baby, when awake, is busy exploring different movements, one after the other. In the beginning, these movements are fumbling, but after some practice, they become more accomplished until the baby switches to another movement. If outer and inner requirements are sufficient, the baby will learn to make these movements in an exact way before starting to make a new one. In order to integrate primitive reflexes completely, the baby must learn to do certain movements in an exact way. Sometimes there are inner obstacles to making certain movements (e.g., low muscle tone and an inability to lift the head). Sometimes adults may restrict the baby's ability to move freely—for example, by not allowing the baby to spend time on the floor.

The baby follows an inner program of motor development and reflex integration. When the baby is allowed to spend time on the stomach, he will exercise and increase the strength of the extensor muscles of the back and the neck. Usually about a month after birth, while lying in a prone position, the baby will start to integrate the tonic labyrinthine reflex by lifting his head. If the baby's muscle tone is so weak that he cannot lift his head in a prone position, there may be insufficient vestibular stimulation of the brain stem, causing the muscle tone to remain low and jeopardizing motor development.

Thanks to the stimulation of the brain stem from the baby's movements, the strength of the extensor muscles of the back of the body increases, causing the chest and legs to rise automatically when the head is raised. When this happens, the Landau reflex has developed.

Rolling the bottom from side to side in a prone position will prepare the baby for crawling on his stomach. Many babies do not crawl on the stomach, but those who do it in an exact way will integrate the crawling reflex and start to integrate the Babinski reflex as well. Before she is six months, the baby should have developed the rolling reflexes and be able to roll from the supine to prone positions, and vice versa.

Between six and nine months of age, the symmetrical tonic neck reflex (STNR) is developed, which assists the baby in being able to sit up. When the head is bent backward, the legs are flexed and the arms are stretched. When the head is bent forward, the arms are flexed and the legs are stretched. The baby now will start to rock back and forth with slightly bent arms. In this way, the STNR is integrated, and not until the baby has made these movements for a sufficient time will he be able to crawl on hands and knees.

By crawling on hands and knees, the baby exercises cross movements, balance, and stability. In this way, the baby will have a better basis for rising, holding on, and moving along furniture and then starting to walk.

The Laundau reflex and the STNR are transitional reflexes that are developed after birth in order to assist the motor development of the baby. These reflexes should be integrated or else they will obstruct the motor development of the child. The STNR reflex is not integrated unless the baby learns to rock back and forth in an exact way.

Problems with linking up of the basal ganglia

If the spontaneous movements of the baby are obstructed, the primitive reflexes will remain active and the linking up of the basal ganglia will suffer. The result can be ADHD, learning difficulties, CP, and other conditions.

There may be many reasons why the baby does not manage to make the rhythmic infant movements satisfactorily. In cerebral palsy, usually a brain lesion has occurred during pregnancy or at birth. In other cases, the baby may have contracted a serious disease at some crucial stage of his motor development. Some babies seem to be in a hurry; they develop early, skip stages like crawling on hands and knees, and learn to walk before ten months of age. The cause is often hereditary. Other children are late to develop; they have low muscle tone and are sluggish and do not like to move; and they need stimulation and time before going on to the next stage.

The effects of restricting the baby's freedom of movement

It is natural for parents to want to help their children develop their motor abilities. When it comes to emotional development, parents are guided by instincts and their own childhood emotional experience. However, when it comes to promoting motor development of their children, parents are usually not aware of what babies need for their development.

Instead of letting the baby develop his motor abilities on his own and at his own pace and letting him rise from a lying to crawling and finally standing position, most parents try to rush motor development by putting him into car seats, strollers, or baby walkers long before he is even able to sit on his own.

All this will restrict the natural motor development and obstruct the linking up of the basal ganglia and the integration of primitive reflexes.

If parents want to help their children develop their motor abilities, they must start from the level of development of their babies. A baby that cannot sit is better off lying. An infant must be allowed to lie on his stomach in order to be stimulated to lift his head. He must be encouraged to move around freely on the floor; to develop his movement patterns gradually; to master gravity, balance, and stability; and integrate his primitive reflexes. By letting the baby spend a lot of time in baby seats and baby walkers, his motor development will be obstructed and he will run a greater risk of developing ADHD or ADD, learning difficulties, and emotional problems.

Damage of the basal ganglia due to environmental factors

As stated previously, research in Sweden has shown that irradiation from a switched-on mobile phone can damage the basal ganglia in young rats within two hours. Considering the fact that pregnant mothers may talk on their mobile phones hundreds of hours during pregnancy, one should not be surprised at the increasing rate of attention problems and hyperactivity in children nowadays.

Mercury is one of the most toxic substances in existence, and it is especially harmful to the brain and neural system. Mercury from amalgam is known to cause damage of the basal ganglia and Parkinson's in adults. Mercury from the amalgam of the mother will be transmitted to the fetus and damage its brain. In my experience, this is a common cause of problems with attention and learning. According to a study from the Faroe Islands, adverse symptoms of mercury exposure are challenges with memory, attention, language, and visual-spatial perception in children at seven years of age.[46]

Food additives such as MSG and aspartame are also known to damage the brain. Metabolites of aspartame are particularly harmful to fetuses and children and are known to damage the brain. In addition, there are thousands of chemical substances that we know little or nothing about. Some of these are known to affect the development of the fetus in a harmful way.

The effect of rhythmic movement training

As has been shown, infants integrate primitive reflexes and develop lifelong postural reflexes by making spontaneous infant movements according to their inner directions. Rhythmic movement training has been developed based on such spontaneous movements. Rhythmic exercises stimulate the development of the basal ganglia and their linking up with other parts of the brain. The rhythmic exercises will integrate primitive reflexes, develop postural reflexes, and improve the child's ability to sit still and to automatize movements.

In addition, the rhythmic exercises affect the brain in many other ways, as has been described.

Examining of primitive reflexes

In infants, primitive reflexes are easy to test. The only thing you have to do is to stimulate the child in a way that will trigger the response. Different reflexes are triggered by different stimulation of the tactile, proprioceptive, vestibular, visual, and auditory senses.

For instance, the tonic labyrinth reflex (TLR) can be triggered by gently bending the head of the baby forward or backward. When the head is bent forward, the baby will go into fetus position and when it is bent backward the baby will straighten his body.

In an adult person, this reflex can be tested by asking the client to stand in upright position with her feet together, close her eyes, and bend her head backward or forward. If the TLR is not integrated you may notice that the test person loses her balance or starts swaying when the head is bent. The person with a slightly active TLR may be able to compensate for this reaction voluntarily so it will be hard to discover. If the test is repeated several times or is done for long periods, the test person will tire of compensating and will eventually lose her balance.

If, on the other hand, the TLR has been integrated into the total movement pattern of the test person, the bending of the head will not cause the person to lose her balance. This means that the bending of the neck has been made automatic on the level of the basal ganglia and that there is no need for compensation.

If it is difficult to decide if the test person is compensating or not, you do not have to repeat the reflex test several times. It is also possible to decide if the reflex is active or not by muscle monitoring.

Integration of primitive reflexes

Rhythmic exercises to integrate primitive reflexes are an excellent method for children and can be used for adults as well. The same exercise may integrate several reflexes, which is very practical. However, for older children and especially adults, it may sometimes be a slow and trying procedure only to work with rhythmic exercises. In such cases, it could be helpful to supplement the rhythmic exercises with alternative ways of integrating primitive reflexes.

British psychologist Peter Blythe has developed exercises that are similar to the reflex integration movements that infants spontaneously make. However, these movements lack rhythmical elements and therefore do not have the same impact on the RAS, muscle tone, and the cerebellum as the rhythmic movements have. Although these exercises are helpful to integrate primitive reflexes, in my experience, reflexes are integrated more rapidly by the rhythmic exercises.

A more effective way of integrating primitive reflexes is taught by Svetlana Masgutova. The principle of this treatment is to reinforce the reflex pattern with a slight isometric pressure. The client is asked to remain in a position that imitates the pattern of the primitive reflex, and the therapist reinforces this movement pattern with an isometric pressure in various directions. The pressure should be held for at least six to seven seconds while the client exhales and repeated three to seven times in order for this treatment to be effective. Usually such an integration needs to be repeated for a number of times for the reflex to remain integrated.[47]

Chapter 8

Primitive Reflexes Especially Important in ADHD

ADHD and retained primitive reflexes

Children diagnosed with ADHD always have retained infant reflexes that affect some of the following capabilities: posture, muscle tone, and the ability to sit still and sort out irrelevant impressions.

Cooperation between the back and front of the body is dependent on our ability to rise and straighten the back. In order to develop this cooperation, the baby must learn to raise the head and chest in a prone position, straighten the back, get up on hands and feet, and finally rise and walk. Certain primitive reflexes have a basic importance for this development and for adequate muscle tone of the extensor muscles (i.e., the muscles that straighten the body). Among others, these are the tonic labyrinthine reflex, the Laundau reflex, and the symmetrical tonic neck reflex. Retained reflexes that cause low muscle of the extensor muscles and poor posture may cause attention problems due to insufficient arousal of the cerebral cortex by the reticular activating system (RAS).

Hyperactive children who have problems sitting still and always seem to fidget often have a retained spinal galant and/or spinal Pereze reflex. Children who are easily disturbed and have problems sorting out irrelevant impressions usually suffer from retained stress reflexes: the fear paralysis reflex and/or the Moro reflex.

The following survey of the most important primitive reflexes not only deals with reflexes usually active in ADHD but also in learning difficulties and in motor problems.

1. The Tonic Labyrinthine Reflex (TLR)

In the womb, the fetus lies in a fetal position with its head bent forward and her arms and legs bent. This is the position of the tonic labyrinthine reflex forward, in which the pattern is that the trunk, arms, and legs are bent when the head bends forward.

Fig. 6 The tonic labyrinthine reflex forward

This reflex develops twelve weeks after conception and should be integrated three or four months after delivery. The tonic labyrinthine reflex backward is developed at the time of delivery. In TLR backward, the whole body is extended and the tonus of neck, back, and leg extensors is increased. TLR backward should be integrated by the age of three years.

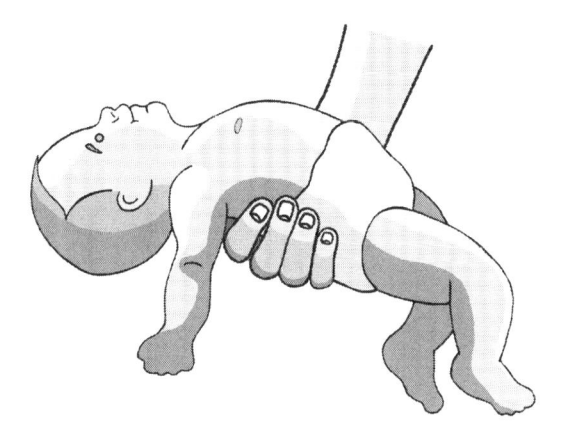

Fig. 7 The tonic labyrinthine reflex backward

TLR helps the child adapt to the new gravitational conditions after delivery and gives the child an early primitive reaction to the gravitational force. Every bending of the head forward decreases the tone of the extensor muscles and neck; back and legs are bent. Every bending of the head backward increases the tone of the extensors, and the body is stretched. The proprioceptive sense is stimulated by the change of muscle tone, and the reflex gives the child an opportunity to practice balance, muscle tone, and proprioception.

If the TLR is not integrated, the effect will be that every head movement backward or forward changes the muscle tone and confuses the balance center. These children have difficulties judging space, distance, depth, and speed.

Symptoms of an active TLR

Children with an active TLR forward can have the following problems:

- Difficulties in holding the head up—it may lean forward or to the side
- Weak neck muscles

- Shrunken posture
- Weak muscle tone, over-flexible joints
- Problems lifting the arms or problems climbing
- Problems with the working of eye muscle—a tendency to be cross-eyed
- Balance problems, especially when looking downward

Children with an active TLR backward can have the following problems:

- Tense muscles, a tendency for toe walking
- Problems with balance, especially when looking upward
- Coordination problems

If the TLR has not been integrated in childhood, there will be other retained reflexes. In adults who have integrated the TLR in childhood, the reflex may be reactivated because of an injury of the neck, the head, or even of the back. The symptoms will be balance problems and pain in the back of the neck.

Importance of the TLR in ADHD and ADD

The tonic labyrinthine reflex is usually active in ADHD and ADD, especially children with ADD, and they often have low muscle tone and poor posture due to a retained TLR forward. There is insufficient stimulation through the reticular activation system to the cortex and especially the prefrontal cortex, causing problems with attention and concentration.

If the fear paralysis reflex is not integrated, the TLR may be impossible to integrate permanently. If the TLR backward and the fear paralysis reflex are both active, there may be hypertonic muscles of the back of the body, causing the child to become a toe walker.

2. The Landau Reflex

At the age of four weeks, an infant in a prone position starts to lift his head from the bed. After another month or two, the child will also lift his chest when the head is raised (upper Landau). After the age of four months, the child will start to extend his legs so they are raised from the bed while lifting his head and chest (lower Landau). The Landau reflex should be integrated at the age of three. When the reflex is integrated, the child in a prone position will keep his legs on the floor while lifting his head.

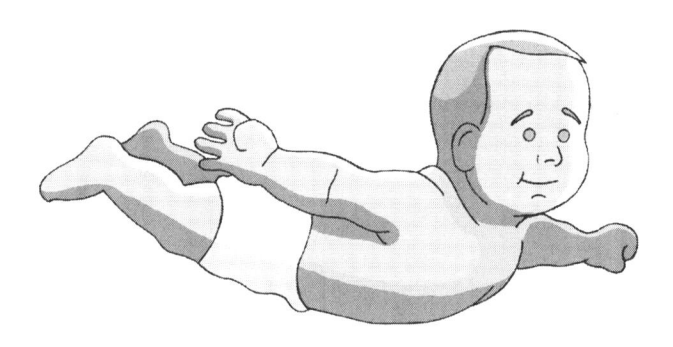

Fig. 8 The Laundau reflex

The Landau is important for the integration of the TLR forward and helps to increase muscle tone in the back and neck in a prone position. When the infant is able to lift his chest from the bed, his arms become free so he can grasp things and bring them to his mouth. This helps with the development of near vision. The lifting of his head and chest also gives him a general view of the environment and opportunity to practice his three-dimensional vision.

If the Landau does not develop properly, the child will have low muscle tone, especially in the neck and back, and will have difficulties raising his head and chest in the prone position. There may be problems doing the breaststroke.

If the Landau is developed but not integrated, the child may have tensions in the back of the legs and become a toe walker. The knees may be extended backward. Later in life there may be pain of the knees and eventually osteoarthritis. Efficient cooperation between the upper and the lower part of the body will be difficult since the legs are extended when the head is bent backward. If the Landau reflex is not integrated, it may obstruct the integration of the spinal galant reflex.

The importance of the Landau reflex in ADHD and ADD

If the Landau reflex has not developed adequately, there will be low muscle tone of the back of the body and poor posture. This may cause insufficient stimulation through the reticular activation system to the cortex and especially the prefrontal cortex, causing problems with attention and concentration. If the spinal galant cannot be integrated due to a retained Landau reflex, there may be problems with hyperactivity and sometimes bed-wetting.

3. The Symmetrical Tonic Neck Reflex (STNR)

The STNR develops when the baby is about six months old and should have a short life span. Therefore, like the Laundau reflex, it is not a genuine postural reflex. It should be integrated by the age of nine to eleven months.

The reflex pattern of the STNR is the following: when the child kneels on hands and knees, the arms are extended and the legs are flexed when the head is bent backward. When the head is bent forward, the arms are flexed and the legs are extended.

The Rhythmic Movement Method

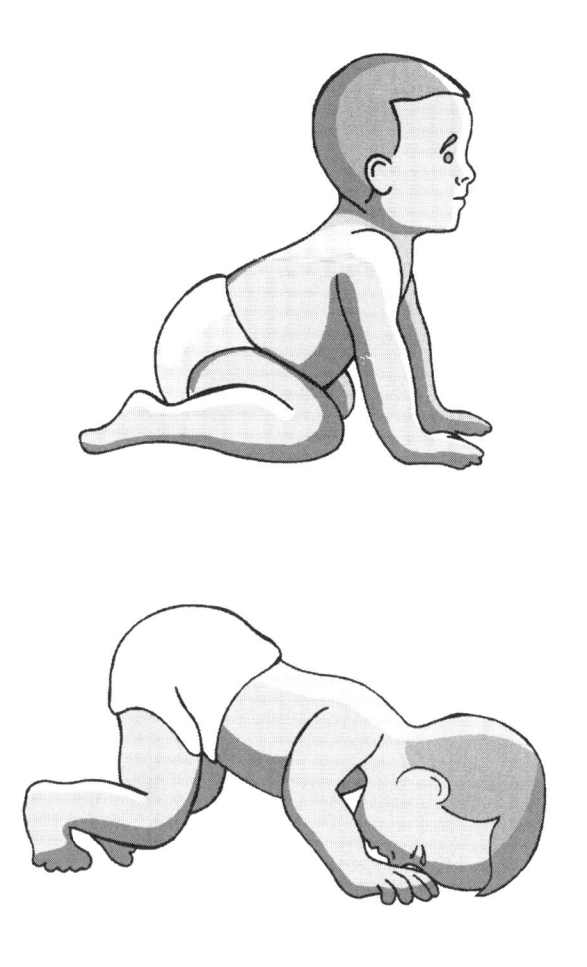

Fig. 9 The symmetrical tonic neck reflex (STNR)

Integration of the STNR helps to strengthen the muscle tone of the back of the neck and the back and is important for proper body posture. Thanks to the STNR, the child can rise on hands and knees from the prone position. But before the child learns to crawl on all fours, the STNR must be integrated, at least to a certain degree, in order for arms and legs not to be dependent on the position of the head. If the STNR is not sufficiently integrated, the child will move around by sliding on his bottom or just sit until he rises and walks. Children who never crawled on all fours usually have an active STNR.

The integration of the STNR takes place when the baby kneels on all fours and rocks back and forth. The STNR influences body posture and strength in the upper arms. The STNR helps the child to train his vision to focus at short and long distances.

A nonintegrated STNR causes bad body posture. The child sits like a sack of potatoes. When sitting at a table, reading, or writing, the child often ends up lying over the book. To prevent this, the child will often support his head with his hand. In order to keep an erect body posture, the child may prefer to fold his legs under himself and get into a W position. Children with a retained STNR may have difficulties with accommodation and focusing at short and long distances, which may cause problems in ball games when the child must follow the ball with his eyes. Children with a retained STNR may have problems with binocular vision. Problems with accommodation and binocular vision are frequent in reading difficulties, and the STNR reflex is usually retained in this condition.

In children and adults with rotated pelvises, the STNR is often active. If the reflex is integrated with isometric pressure, the rotation usually decreases or disappears, depending on which other reflexes are involved.

A common symptom of the STNR is weakness of the upper arms, and the child may have problems doing a somersault. The child will also often have challenges doing the breaststroke because of difficulties bending the head backward and bending the arms and stretching the legs at the same time. Because of poor coordination of upper and lower parts of the body, the lower part may have a tendency to drop down when swimming.

Importance of the STNR in ADHD

A nonintegrated STNR will cause bad posture and problems keeping the back straight. When the child looks down, she will shrink up and lean over the desk, which will obstruct breathing. There may be insufficient stimulation through the RAS to the neocortex and prefrontal cortex, especially if the child is bored or in stress, resulting in problems with attention and concentration.

4. The Spinal Galant Reflex

If you touch the area next to the spine on either side at the level of the waist, the hips of the client will turn toward the side being touched.

Fig. 10 The spinal galant reflex

The reflex develops twenty weeks after conception and should normally be integrated three to nine months after delivery. This reflex is considered important for the conducting of body vibrations in the fetus and the development of the vestibular system. The reflex also helps the baby move down the birth canal during delivery. If this reflex is not integrated, the development of the amphibian reflex is impaired, which can cause clumsiness of the lower part of the body and tension of the legs.

Children with a nonintegrated spinal galant reflex are often restless and hyperactive. Tight clothing or belts or just leaning against the back of a chair can trigger the reflex and cause the child to fidget. Most children who have not integrated this reflex prefer to wear loose clothes. Some children with an active spinal galant reflex are bed wetters. If the reflex is active only on one side, there can be scoliosis of the spine.

Persons with an active spinal galant reflex sometimes learn to fixate the lumbar spine, which may cause back problems. Fixation and rigidity of the spine at this level impairs the cooperation between the upper and lower body and may cause problems in getting in touch with feelings. In adults, a retained reflex may cause lower back pain, and a rotated pelvis is common in both children and adults.

5. The Spinal Pereze Reflex

The Spinal Pereze reflex is a primitive reflex that emerges at birth and is integrated between three and six months after delivery. Stroking with a finger along the spine from the tailbone to the neck causes the baby to lift his head and bottom, bend the thoracic spine backward, and bend arms and legs.

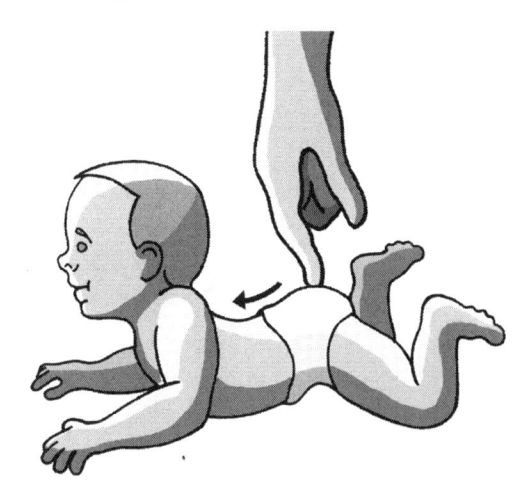

Fig. 11 The Spinal Pereze reflex

The Spinal Pereze reflex assists the development of the Laundau reflex and the STNR and helps the baby to get up on hands and knees between the age of six to nine months.

Symptoms of a nonintegrated Spinal Pereze reflex

Delayed development of this reflex may cause a lack of muscle tone in the back and general hypotension. A retained reflex may cause sensitivity and muscle tensions, especially of the thoracic back. It may also cause a rotated pelvis. Sometimes adults may have lower back pain and tensions of the legs.

Persons with a retained Spinal Pereze reflex may be very sensitive in the thoracic back and feel ill at ease having people behind them. They prefer to sit against a wall or in the back. They may also have problems sleeping on their backs.

Moreover, the symptoms are the same as with an active spinal galant reflex: restlessness and sometimes bed-wetting in children.

6. The Amphibian Reflex

The amphibian reflex is a lifelong postural reflex that develops when the child is between four and six months old.

Fig. 12 The amphibian reflex

Raising the pelvis on one side causes an automatic flexion of the arm, hip, and knee on the same side. The reflex is first developed in the prone position and then in the supine.

The amphibian reflex starts developing only when a certain amount of integration of the ATNR has taken place and the movements of the arms and legs no longer are totally dependent on the position of the head. Before the child learns to cross crawl, the amphibian reflex must have developed. This reflex helps the child to bend his legs and get up on hands and knees. The development of the amphibian reflex helps to integrate the spinal galant reflex. If the amphibian reflex fails to develop, it signifies that the Spinal Galant and possibly the ATNR reflexes have not been integrated.

Adults who have not developed their amphibian reflex often suffer from clumsiness in the lower part of their body and tension in their legs.

The importance of stress reflexes in ADHD, emotional problems, and reading difficulties

The fear paralysis reflex and the Moro reflex are usually retained in children with challenges. Due to the high levels of environmental stress in modern society, many children do not integrate these reflexes. Children with these reflexes active are hypersensitive to sensory stimulation and easily distracted. Their stress level is high, which will obstruct concentration and learning. Emotional symptoms are common when these reflexes are not integrated. In reading difficulties, these reflexes may cause visual problems that can seriously obstruct reading.

7. The Fear Paralysis Reflex

The fear paralysis reflex (FP reflex) is one of the early withdrawal reflexes that emerge in the second month after conception. The withdrawal reflexes are characterized by a rapid amebic-like withdrawal movement as a response to a tactile stimulation of the mouth region.[48] The pattern of the FP reflex has been described as a terrified rabbit, completely frozen on the spot and unable to move. The reflex should normally be inhibited before the twelfth week after conception and integrated into the Moro reflex. If the FP reflex is not inhibited, the Moro reflex will usually stay active, and in many cases, the TLR will also stay active. Balance problems are therefore common.

The FP reflex is not a primitive reflex since it is not triggered by way of the senses, which have not yet developed. The reflex should be regarded as a reaction to stress by the cells of the fetus. Unicellular organisms react to stress by moving away from the source, whether it is toxicity or some other danger. But the cells of the fetus have nowhere to go, and instead they protect themselves from the environment by producing stress proteins that make the cell membranes less penetrable and diminish the active transport over the cell membrane. At the same time, the fetus is paralyzed and stops moving.

When the FP reflex emerges, the neural system is not sufficiently developed to be able to transmit the reflex pattern by way of neural impulses. Instead, the information is transmitted directly between the cells by way of electromagnetic frequencies. Stressful factors that prevent the integration of the reflex may be electromagnetic irradiation, heavy metals, and other toxic substances. Also, stress from the mother can trigger the reflex and prevent its integration. If there are extremely stressful conditions for the fetus during the first months after conception, the reflex may be constantly triggered and the fetus may be frozen in immobility most of the time, preventing the reflex from integrating.

Children and adults with an active FP reflex have a low tolerance to stress. There is oversensitivity to the senses, usually touch, sound, light, or sudden changes of the visual field and vestibular and proprioceptive stimulation. Sometimes there may be oversensitivity to smell and taste. Persons with a strong vestibular sensitivity and a disposition to motion sickness may feel dizzy and sick when they do rhythmic exercises involving the head, and these exercises may actually trigger the reflex in some cases. When the reflex is triggered, there is a release of the stress hormones cortisol and adrenalin. In adults, panic syndromes and social phobias may occur—and sometimes high blood pressure. Muscle tension in the neck and shoulder areas is common.

It is usually stressful for children and adults with an active FP reflex to look another person in the eyes and a common symptom is difficulty meeting the eyes of another person. Some persons have learned to compensate for that by staring intensely into people's eyes, often without blinking.

Testing and integration of the fear paralysis reflex

To test the reflex, you can look the person in the eyes and walk toward him. Children will often look away. Adults and children who have learned to compensate for the reflex may meet your eyes without being consciously aware of emotional stress, but they usually show some signs of discomfort like tensing up, biting their lips, or smiling.

Adults and some children with an active FP reflex may be difficult to muscle test because they easily react with hypertonic muscles when the reflex is triggered. Some persons with retained FP reflexes may not be able to lock their muscles in muscle testing.

The fear paralysis reflex is a cell reaction to stress and can be triggered any time by an excess of environmental stress, even if the reflex has been integrated in the womb. In order to integrate the fear paralysis reflex, you need to eliminate factors that trigger it. It is especially important to diminish environmental stress by protecting the child from electromagnetic irradiation. Excluding harmful food additives like MSG and aspartame is crucial. In cases of intolerance to casein and gluten, these should be excluded from the diet.

The FP reflex can be integrated both by passive rhythmic exercises and by movements that are similar to movements the fetus makes during the first months after conception. These so-called pre-birth exercises have been proposed by Claire Hocking, and in my experience, these exercises may integrate the reflex if they are done a few times a week for at least a month, usually for much longer. The fact that these exercises are effective indicates that similar early movements are important to integrate the reflex of the fetus.

The pre-birth exercises are surprisingly powerful and may initially activate the reflex and worsen the symptoms in sensitive children, especially those with food intolerance or exposed to high levels of environmental stress (e.g., heavy metals, electromagnetic irradiation, and so on). In such cases, the exercises should be interrupted and the environmental stress diminished before continuing the exercises. Experience shows that in most cases, the reflex cannot be integrated by the pre-birth exercises if there is extensive stress from food intolerance or electromagnetic radiation.

8. The Moro Reflex

The Moro reflex starts to develop after the twelfth week, should be fully emerged in the thirtieth week in utero, and should normally be integrated about four months after delivery. When the FP reflex is not inhibited, the development and subsequent integration of the Moro reflex will be obstructed. Therefore, the Moro reflex is usually active when the FP reflex is not integrated.

The Moro reflex is triggered by a strong and unpleasant stimulation of the balance, auditory, visual, tactile, or proprioceptive sense—for instance, a sudden change of the position of the head, a loud sound, a frightening visual stimulus, an unpleasant touch, or a sudden change of position. The infant reacts in a characteristic way: first by taking a deep breath and stretching his arms and legs out away from the body, and then the arms and legs are bent into the middle of the body and the infant starts to cry.

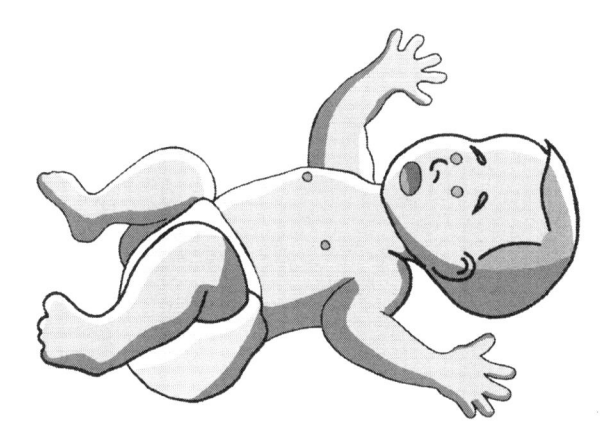

Fig. 13 The Moro reflex

In the womb, the Moro reflex movement will help the fetus exercise its respiratory muscles. When the midwife triggers the Moro reflex in the newborn baby to initiate breathing (e.g., by letting the head fall slightly backward), she will trigger the Moro response and the baby will start crying. In premature babies born before the thirtieth week, this response cannot be elicited because the reflex is not yet fully developed.

When the Moro reflex is activated, the defense mechanisms of the body are alerted. The sympathetic nervous system and the adrenals are stimulated and the stress hormones epinephrine and cortisol are secreted. Epinephrine causes the senses to become oversensitive. A nonintegrated Moro reflex causes many different symptoms from one or more senses:

- Visual sense: big pupils that are slow to react to light, which causes bad twilight vision and hypersensitivity to light
- A tendency to be cross-eyed at both near and far distance
- Auditory sense: hypersensitivity to sound or specific sounds, difficulties shutting out back ground noise
- Vestibular sense: hypersensitivity to vestibular stimulation, motion sickness, problems with balance

- Tactile sense: hypersensitivity to touch
- Kinesthetic sense: hypersensitivity to sudden change of position.

Both an integrated Moro reflex and FP reflex may be activated in situations of extensive physical, emotional, or environmental stress. These two reflexes are usually activated in cases of burnout and chronic fatigue syndrome in adults.

CHAPTER 9

PRIMITIVE REFLEXES IN READING AND WRITING DIFFICULTIES

Reading and writing difficulties and the midline reflexes

All children with reading and writing difficulties have retained primitive reflexes that may obstruct the cooperation of the hemispheres of the brain and the cooperation between the eyes or binocular vision. The movements of the newborn infant are homolateral (i.e., the baby moves one body half independently of the other). The cooperation of the hemispheres is slight, and the nerve connections between the hemispheres through the corpus callosum are undeveloped.

When the child learns to move in a way that involves both hemispheres simultaneously, so-called cross movements, the nerve connections of the corpus callosum will be stimulated. For instance, such movements are moving objects from one hand to the other and crawling on the stomach or on hands and knees.

The left and right hemispheres are specialized in different functions. The left hemisphere is in charge of language abilities like analyzing the sounds of the language and speaking and understanding spoken language. The right hemisphere is more specialized in emotional undertones and understanding context. Even if the language and speech centers are located in the left hemisphere, that does not mean that the left hemisphere is more important than the right one when we read. Brain scans have shown that both hemispheres are equally activated. Good reading ability is therefore dependent on a working exchange of information between the hemispheres through the corpus callosum.

Not until the child integrates the reflexes that, if non-integrated, could prevent the cooperation between the hemispheres is a proper exchange of information between the hemispheres made possible. These are the so-called midline reflexes—reflexes that affect both sides of the body. Most important of these reflexes is the asymmetrical tonic neck reflex (ATNR), but several other reflexes of the hands and feet are also important.

9. The Asymmetric Tonic Neck Reflex (ATNR)

The asymmetric tonic neck reflex (ATNR) develops about eighteen weeks after conception and should be integrated when the baby is about six months old.

When the baby turns his head to one side, the arm and leg on that side are stretched, while they are bent on the other side.

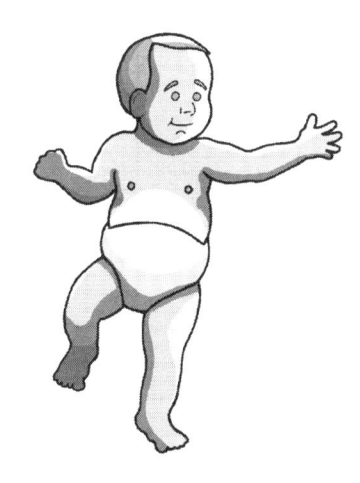

Fig. 14 The ATNR reflex

In the fetus, the ATNR triggers kicking movements and gives proprioceptive and tactile stimulation. The reflex assists the child during delivery. It causes the newborn child to move his arms and legs, depending on the position of the head. These movements are homolateral (i.e., the child moves one body half separately and stimulates the left and right hemisphere separately). The gradual modification of these movements assists the child in integrating the reflex: lying on his back and bending his arms or legs and tracking the fingers or toes with his eyes and eventually putting them in his mouth is one way of integration. Lying in a prone position and lifting his head and chest, grasping things and putting them in his mouth, is another. Further integration takes place by cross movements causing nerve signals to pass through the corpus callosum and stimulating its nerve connections.

The movements that the infant makes to integrate this reflex also train his binocular vision (the ability of the eyes to cooperate) and the ability to track moving objects with his eyes.

If this reflex is not integrated, the child may have difficulties making cross movements and have difficulties crossing the midline. The child may walk in a slow amble. The balance of the child is affected when the head is turned to the side, making it difficult for him to learn to ride a bicycle. When the child turns his head to the right, the right arm and fingers are extended and the child easily drops things or turns them over. When writing, the child compensates for this by gripping and pressing down hard on the pen or pencil, which impairs his handwriting. Some may have challenges writing eights.

A retained ATNR may cause visual problems such as deficient binocular vision, astigmatism, and sometimes strabismus and problems with tracking. Adults may have the same visual problems. More often, there are tensions and pain of the back of the neck, shoulders, back and hips.

10. The Babkin Reflex

Lightly pressing on the palms of a baby triggers the Babkin reflex. Then he opens his mouth and bends his head forward or to the side and starts doing sucking movements with the mouth. When the baby sucks, you can also observe involuntary movements of the hands.

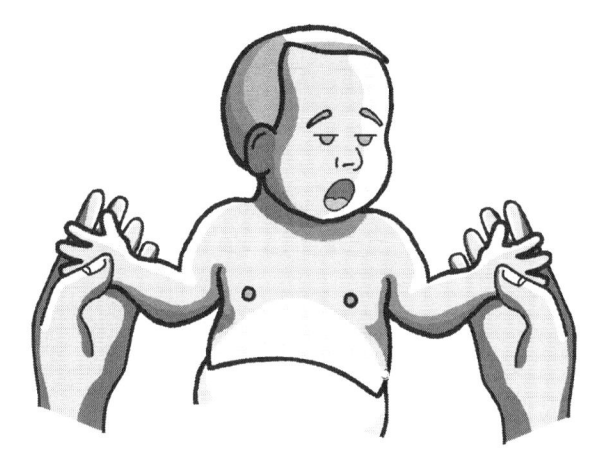

Fig. 15 The Babkin reflex

The movements of the hands may stimulate the breasts when the baby is breast-feeding. The reflex will assist the baby in putting his thumb or things into his mouth.

The Babkin reflex is developed in the second month after conception and is active during the first three or four months after delivery.

If the Babkin reflex is not integrated, the child will have challenges with the motor control of his hands. The fingers may be very floppy. Fine motor skills will be impaired, with challenges in tying shoelaces, doing up buttons, and so forth. Poor handwriting can occur. There may be challenges with speech, articulation, and often involuntary movements of the mouth and tongue when the child writes, plays an instrument, or when using scissors. Adults with a retained Babkin reflex often have tensions of the jaws and may bite or grind their teeth at night. Children with this reflex active may also have a tendency to bite or chew pens, clothes, and so on.

Due to the difficulties of articulation, the corresponding area of the sensory cortex of the left parietal lobe is not properly stimulated, which may cause impaired phonological ability and difficulties perceiving sounds. By motor training and integration of the Babkin reflex, articulation will improve, as will phonological ability.

11. The Grasp Reflex

Putting a finger in the hand of a baby triggers the grasp reflex. The baby will grasp the finger and hold it. If you lift the baby, her arms will stretch.

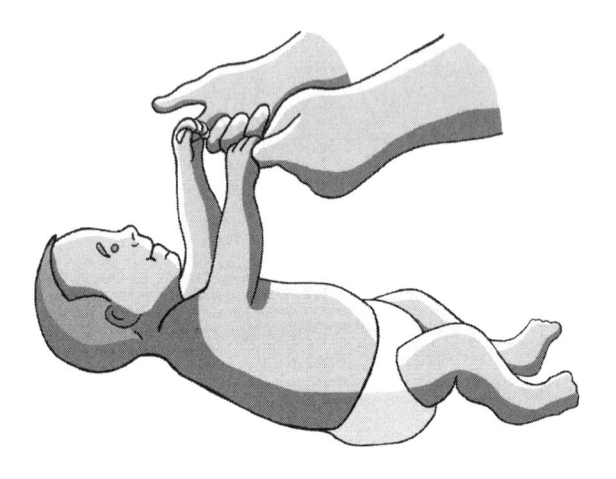

Fig. 16 The grasp reflex

The grasp reflex is developed the third month after conception and should be integrated during the first year after delivery. The reflex is important for hand-eye coordination, development of binocular vision, and cooperation between the hemispheres. The baby grasps things and looks at the object, bringing it to the mouth. Later the reflex is important for ear coordination and the ability of the auditory sense to judge distance and direction. When the baby has learned to support herself sitting in a child's chair, she will start to throw objects around, practicing throwing and releasing things at the same time. In this way, the baby simultaneously integrates the grasp reflex and learns to judge direction and distance of the sound when the object hits the floor.

If the grasp reflex is not integrated, the child may have challenges with the motor control of his hands, poor handwriting, and poor fine motor skills. The pencil grip will be poor or unusual, and there is a tendency to hold the pen too tightly. An active grasp reflex causes tension in the shoulders and makes writing difficult. In adults, there are often tensions of the shoulders and problems differentiating the movements of the hands and the shoulders. One effect may be difficulties holding a firm grip of a golf club while making a swing, resulting in throwing away the golf club with the ball.

12. The Hands Pulling Reflex

The hands pulling reflex is triggered by holding the baby around his wrists and pulling him toward you. Then the baby bends his arms and helps to get up into a sitting position.

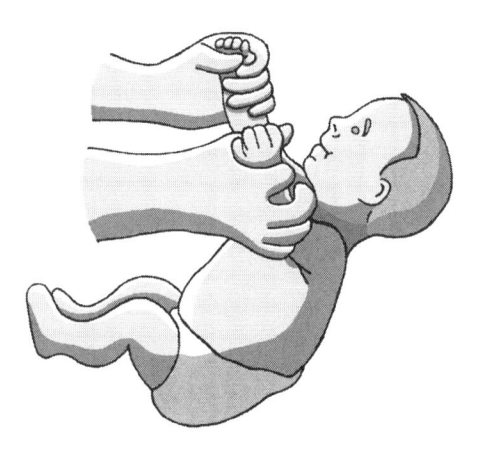

Fig. 17 The hands pulling reflex

This reflex emerges twenty-eight weeks after conception and is normally integrated two to five months after delivery.

At the age of two months, the grasp reflex is merged with the hands pulling reflex and they start to function as a unit: when you put your fingers into the palms of the baby, she will clutch your fingers and bend her arms so you can help her get up a into sitting position. These two reflexes enable the child to learn how to handle objects with her hands, pulling them toward her, putting them into her mouth, throwing them away, and so forth. They also assist the integration of the Babkin reflex.

An active hands pulling reflex causes tensions in the forearms, making writing difficult. In some persons, the arms are constantly bent at the elbows; others have problems keeping them bent. In adults, the tensions of the forearms may cause problems of the elbow (e.g., tennis elbow). In children who are in the habit of flapping the forearms when they get excited, this reflex is usually active.

13. The Babinski Reflex

The Babinski reflex develops during the first month after delivery and should be integrated at the age of two.

When you stroke with a pen along the outer lateral part of the sole of the foot from the heel to the little toe, the big toe is extended and the other toes spread.

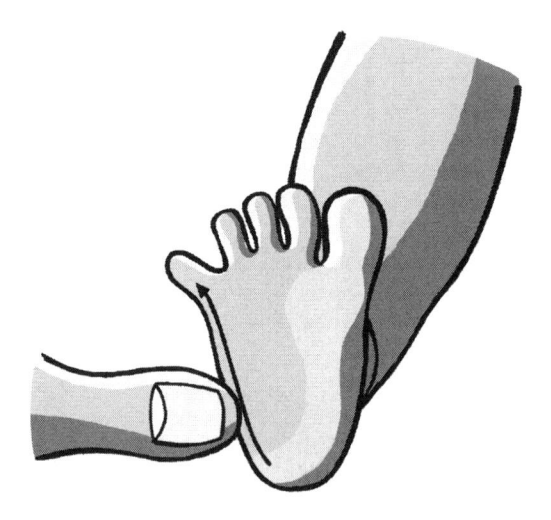

Fig.18 The Babinski reflex

The Babinski reflex is important in preparing the feet for walking and influences the ability not only to move the feet but also the legs, hips, and lumbar spine. It is important for the muscle tone of the lower part of the body. Children with an undeveloped Babinski reflex are often flat-footed, slow, and do not like walking. They walk on the insides of their feet, and consequently their shoes are worn out on the inside. They may have loose ankles that are easily sprained.

If the reflex is developed but not integrated, the children have a tendency to walk on the outside of the feet. Their shoes are worn out on the outside. These children develop tension in their feet and legs when they grow older.

14. The Plantar Reflex

If you press your thumb against the sole of the baby's foot, between the toes and the arch of the foot, the toes will bend. The reflex emerges when the fetus is eleven weeks and should be integrated seven to nine months after birth.

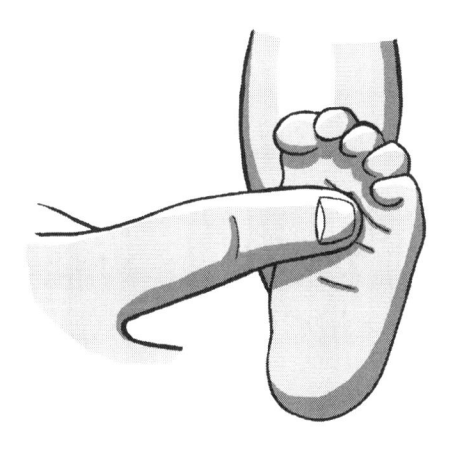

19. The plantar reflex

Like the Babkin reflex, the plantar reflex belongs to a group of grasp reflexes that are considered a remnant from a previous stage of development, when the young ones had to cling to their mothers. In many mammalians, the plantar reflex assists breast-feeding in the same way the Babkin reflex does.

A retained plantar reflex may cause tensions of the jaws and biting and grinding of the teeth, as may a retained Babkin reflex, and therefore affects articulation in a similar way.

A retained plantar reflex may cause phonological challenges and difficulties perceiving sounds in the same way the Babkin reflex does. By motor training and integration of the Babkin reflex, articulation may improve, as will phonological ability.

Like the Babkin reflex, a retained plantar reflex may cause tensions of the jaws and biting.

Usually the Babkin and the plantar reflexes are both retained in children and less often only one of them. If both are retained, both need to be integrated in order to improve articulation and phonological ability.

Case study: Tomas

Tomas was eleven years when his mother contacted me. She was very concerned about him and had tried to get help for his motor problems and his learning challenges. Although a pediatrician had referred him to the rehabilitation clinic, his problems were not considered sufficiently severe to justify a visit there.

Tomas was born after a protracted delivery that lasted for three days. He was late to meet his mother´s eyes and his motor abilities were late to develop. He learned to walk at the age of two and to speak at the age of four. He learned to ride a bicycle when he was eight, but he still had problems riding it. He was too weak to ride it uphill and used to get off the bicycle and walk.

On the whole, he was weak and used to walk very slowly uphill, "as if he were eighty," according to his mother. He never had the strength to carry anything heavy. His mother always had to wait for him when they were out walking. Tomas had not learned to swim or to skate. He was clumsy; he often stumbled and fell. He liked to wrestle but was very rough because he could not adjust his strength and therefore could not help hurting the people he wrestled with. He was restless and had difficulties sitting still. He easily lost his temper and had poor perseverance.

He had great difficulties reading, and an optometrist had prescribed reading glasses. His handwriting was bad, and he wrote with capital letters.

Tomas's first visit

At his first visit, Tomas was dejected. He did not believe he could improve in any way and had no wish to get better at anything. Everything was boring; nothing was fun. He said he had tired of his mother dragging him around to see people who could not help him anyway. He said he no longer would put up with that, adding that he wanted to leave immediately.

I replied that I could understand why he believed he was a hopeless case and that he might be right thinking so. However, I would not be convinced of his being a hopeless case unless he proved it to me.

I offered to show him some exercises to do every day. If he did not notice any positive change doing these exercises for at least a month I would have to agree with him that he was indeed a hopeless case. Tomas reluctantly agreed to my proposal.

Before I showed him the exercises, I examined his vision and reflexes. He had binocular problems with an esophoria at a reading distance that was corrected when he wore his reading glasses. He was very shrunken, with low muscle tone. Several primitive reflexes were nonintegrated, among others the TLR, the spinal galant, and the symmetrical and asymmetrical tonic neck reflexes. His Landau reflex had not developed, and he hardly managed to raise his head in a prone position.

I showed him a few simple rhythmic exercises and told him to do these every day until his next visit after one month. He promised to do the exercises every day.

Tomas's development during rhythmic movement training

A few days before his next visit to me, his mother phoned and told me that Tomas had been doing the exercises every day and that he liked them. She had believed that the exercises were a joke on my part because they seemed so simple, but she had to reconsider because she had already seen great changes in Tomas, so she understood that they were seriously meant.

For more than a year, Tomas continued to do exercises with short intervals almost every day. He visited me every month to adjust the exercises he was doing and learn new ones. He mainly worked with rhythmic movements to integrate reflexes. Not long after he started the training, he got a high fever for a few days. After a couple of months, he had become more cheerful and his endurance had improved. He no longer lost his temper so easily.

After three months, he felt stronger, and his mother had noticed that he was more focused and had fewer problems sitting still. She could reason with him in a new way, and his abstract thinking had started to develop. Over the next few months, he grew stronger and more active. His endurance improved, and he did not get tired as before. He could now ride a bicycle without problems, even fifteen to twenty miles. He could outrun his mother.

After eight months of rhythmic movement training, his reading had improved and he no longer wanted to use his glasses when reading. A visual examination showed that he now had good binocular vision at a reading distance without his glasses; he did not need them anymore. After a little more than a year, he was very proud after having finished the last Harry Potter book. His handwriting was much better. He had learned to skate. He was cheerful and active, and he had good endurance and no problems sitting still.

CHAPTER 10

THE LIMBIC SYSTEM AND RHYTHMIC MOVEMENTS

The limbic system, or the mammalian brain

The limbic system, in the mid area of the brain, developed in mammals—hence it is also called the mammalian brain. The most important task of the limbic system is to regulate emotions.

The development of the limbic system reflects the importance of emotions for social organization and for the survival of the species among mammals, in contrast to the relative unimportance of emotions among reptiles.

Reptiles lay eggs and, apart from the crocodile, do not take care of their young ones, who must manage completely on their own after the eggs have hatched. Some species even need to be careful not to be eaten by their parents.

Unlike reptiles, mammals are viviparous. Their young ones are helpless and dependent on being taken care of until they are ready to manage on their own.

Mammals developed mammary glands to nurse their young. They learned to protect their offspring, react when they called, and come to their assistance when they were hungry, cold, or threatened by outside danger. The care of the parents or the mother became a precondition for the survival of the offspring. Maternal affection became a central premise for the survival of the species.

Being able to get food by sucking and keeping warm by close bodily contact with their mothers became a condition for the survival of young ones. Also important for survival was the development of the ability to call for the assistance of their mothers. Family systems developed as the young were born in litters. The young learned many things by sharing the knowledge and experience of their parents. Through play, the young learned social rules and how to get on with their siblings. By playing, they also trained in hunting and other abilities they needed in order to manage and survive adult life.

The function of the limbic system

The limbic system is of vital importance for our survival—firstly, by regulating the inner environment; and secondly, by playing an important part in the relationship of the individual with the surrounding world. It receives information from the world by signals from the tactile, proprioceptive, vestibular, auditory, visual, gustatory, and olfactory senses.

Equally important are the signals that are transmitted by the vagus nerve to the limbic system from sense organs of the inner organs (i.e., stomach, bowels, and heart, giving information about the inner environment).

When we become aware of a threatening situation through our external senses, the limbic system is activated. We can react with fear and escape or with aggression and attack to eliminate the danger. When we reach safety, we can feel calm again.

When the limbic system gets signals from the inner organs that we need food or water, we experience hunger and thirst; then we do something to get food or water. When we are hungry and thirsty, we feel discomfort, which is changed into pleasure when we eat or drink.

You could say that the limbic system receives information and transforms it to emotions, and in certain situations, these emotions have a vital importance for our survival.

However, the limbic system is not only activated by our external and internal sense organs. Our thoughts can also evoke feelings.

We can feel butterflies in the stomach or a pressure in the chest and become worried or depressed when we think of an unpleasant situation. Alternatively, when we think of a pleasant memory, we can feel happy and satisfied.

Overview of the limbic system

All the behaviors that are mentioned above are controlled by the new parts of the brain, which were developed by mammals: the mammalian brain, or the limbic system.

The limbic system obtained its name from the limbic brain, or the cingulate cortex, which is situated like a border (Latin limbus) around the brain stem and the reptilian brain. According to MacLean, the function of the lobus cingulus is to regulate maternal behavior and play. The front part of the temporal lobe and part of the underside of the frontal lobes also belong to the limbic system. Two important structures in the temporal lobe are the amygdala, which lies in the front part of the temporal lobe, and the hippocampus, which begins inside the temporal lobe and goes like an arch under the gyrus cingulus.

The amygdala is of vital importance for the emotions of anger and fear, for defense and attack reactions, and for food intake and sex. The hippocampus is the seat of our episodic memory (i.e., our memories of events we have experienced personally). It is responsible for our experience of individuality. It is essential for learning, as it is important for converting short-term memory into long-term memory. A very important function of the hippocampus is to form the stem cells that develop into new brain cells.

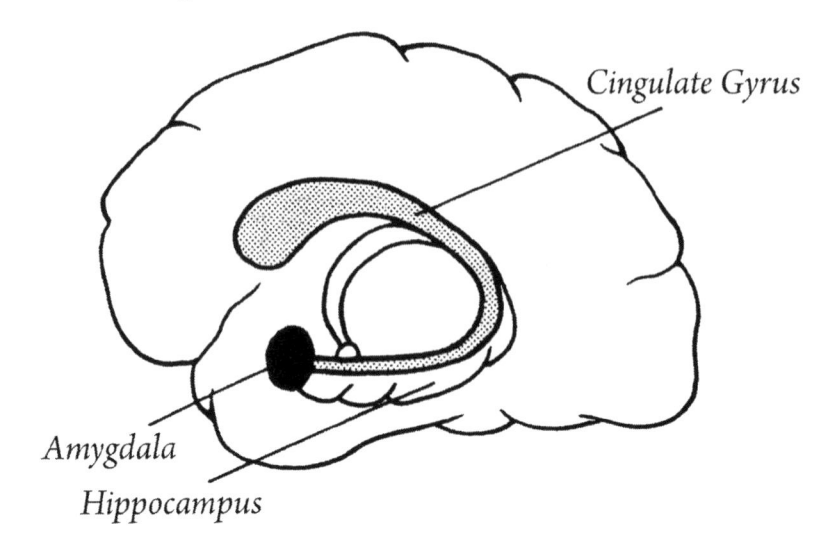

Fig. 20 Overview of the limbic system

Since the limbic system is connected to the important nerve nucleus, the ventral tegmentum, in the midbrain (the upper part of the brain stem), it will be stimulated by the reticular activation system (RAS). From the ventral tegmentum, nerve connections go to the limbic system, ending in the prefrontal cortex. The transmitter substance between the neurons of this nerve pathway is dopamine. The limbic system must function as a unit with the prefrontal cortex in the same way as the motor cortex must function in close cooperation with the basal ganglia. If the limbic system and the prefrontal cortex are not closely connected, our emotions will not be well regulated.

The hypothalamus is closely connected to the limbic system and functions as a controlling center of the inner environment by regulating body temperature, the autonomic nervous system, and the hormone system.

How infants handle stress

Newly hatched reptiles must be able to do everything they need to do in order to survive without any help from their parents. Newborn infants and mammals are completely dependent on their mothers to survive.

When the young of mammals meet an external threat, the Moro reflex is triggered. They cling to their mother if she is around or call for her attention by whining or screaming.

When a newborn baby does not start to breathe after delivery, the midwife can trigger the Moro reflex, causing the baby to take a deep breath and stretch his arms and legs out away from the body. Then the arms and legs are bent into the middle of the body and the infant starts to cry.

The Moro reflex has also been called the clinging reflex and is important for the survival of newborn mammals. By clinging to the mother, the young one seeks protection, and if she is not around, the whining or screaming will get her attention.

When the Moro reflex is triggered, the child usually starts to cry. The child must then be able to cling to or hug an adult in order for the reflex to be integrated. Instinctively parents pick up their crying babies and hug them, rock them, and carry them around until they calm down. In this way, the Moro reflex is integrated.

If the fear paralysis reflex has not been integrated, there may be severe sensitivity to the vestibular and tactile senses, and this reflex may be triggered when the parents hug and rock the baby. Epinephrine will then be secreted, making the baby even more sensitive to tactile and vestibular stimulation. The baby may resist being hugged and rocked, making it impossible for the Moro reflex to integrate.

Tactile and vestibular stimulation develop the emotions

In order for infants to venture out in to the world to explore and interact with the environment they must have adults around who can reassure and calm them when they become stressful and frightened.

In the 1950s, Dr. Harry F Harlow,[49] in his research with monkeys, showed that tactile stimulation is of vital importance for the development of our emotions. He took newborn baby rhesus monkeys from their mothers and raised them with artificial "mothers." There were two groups; in one group, the mothers were plain wire, and in the other group, the mothers were plain wire covered in terry toweling cloth. The monkeys with terry cloth mothers clung to them, hugged them, and got emotionally attached to them. These baby monkeys ventured to explore the surroundings, and when they were frightened, they ran to their mothers and were reassured just by touching them. However, monkeys with mothers made of plain wire did not form any emotional attachment to them and did not develop any sense of security in relationship to their mothers, even if they received milk from them.

For newborn mammals to develop self-confidence, pleasant tactile stimulation is not enough. Vestibular stimulation is also necessary. Mammals get such stimulation when their mothers carry them around. In an experiment similar to the one made by Harlow, one group of baby monkeys was raised by terry cloth mothers who were swinging and rocking, while another group had stationary mothers. Monkeys who were raised by stationary mothers were afraid to explore the environment and overreacted to unfamiliar situations. The other group did not show any such abnormalities.[50]

Emotional challenges due to nonintegrated FP and Moro reflexes

The Moro reflex is developed in the twelfth week in utero, should be fully matured about the thirtieth week, and should normally be integrated about four months after delivery. If the fear paralysis reflex has not been integrated, the Moro reflex will usually remain active. Such a child may live in a constant state of inner stress, oversensitive to external sensory impressions. When the FP and Moro reflexes are activated, the defense mechanisms of the body are alerted. The sympathetic nervous system and the adrenals are stimulated, and the stress hormones epinephrine and cortisol are secreted. Epinephrine causes the senses to become oversensitive.

Due to the condition of constant inner stress, these children may withdraw into themselves and shut off external sensory impressions that they cannot handle. Like the baby monkeys in Harlow's research, they can be afraid of unfamiliar situations and therefore afraid to explore the world. Some of them are extremely shy and may even refuse to speak to anyone except their parents (elective mutism), while others were more likely to act out. Other symptoms may be poor adaptability and inflexibility as well as obsessive-compulsive symptoms. Some children show negativism and oppositional and aggressive behavior. Temper tantrums are common.

Often they have problems in having confident contact with children their own age. They may lack the inner security they need to be spontaneous and flexible and often react with anxiety and outbursts when there is a change of routine and things no longer remain the way they are used to. The lack of emotional security and flexibility can result in a need to manipulate or dominate their playmates. They may become very tired after being exposed to excessive stimuli and may need to rest or even sleep after school.

Some individuals with active fear paralysis and Moro reflexes may be extremely sensitive to the rhythmic movements that stimulate the vestibular system. However, if they perform these movements gently, it is possible that they will be able to make them.

If the fear paralysis reflex is active, it should usually be integrated before the Moro reflex to avoid unnecessary emotional reactions.

Attack and defense behavior and the tendon guard reflex

The Moro reflex is a primitive reflex and as such is controlled from the brain stem. It should be integrated by the age of four months and should be replaced by the fight-or-flight reaction, which, according to MacLean, is directed by the limbic system, or more exactly, by the amygdala.[51]

When we get frightened, blood pressure rises and the blood flow to the skeletal muscles increases. As in the case of the Moro reflex, there is a secretion of stress hormones such as epinephrine and cortisol. By means of the limbic system, these processes trigger emotions of fear and aggression. The fight-or-flight pattern includes processes in the sympathetic nervous system, the hormone system, and the skeletal muscles. When we prepare to fight or flee, we draw breath and hold it. Then the diaphragm is contracted, as are the respiratory muscles of the chest and neck.

The tendon guard reflex is a defense mechanism that is triggered to protect the tendons and muscles from too much tension when the fight-or-flight pattern is activated. When the mechanism activates body posture changes, the flexor muscles of the body are contracted; the calf muscles are shortened, with the knees bent and locked; and the body rises up onto the toes. At the same time, the muscles of the neck and the back are contracted in order to keep the body standing in an erect posture.

We may not want to be aware of feelings of fear and aggression, and then we repress these feelings. However, the body will react anyhow. Persons who live in constant stress may react with symptoms from their inner organs or from their skeletal muscles, and their muscles may generally become too tense.

Wilhelm Reich coined the term muscle armor, describing how we contract our respiratory muscles and diaphragm when we repress feelings of anxiety and anger. When we contract our inspiratory muscles, the chest is expanded and the diaphragm is contracted. The widening of the chest and the secretion of epinephrine may cause hypertension in people who live in constant stress.[52]

The limbic system and memory

As has been pointed out, reptiles are able to do everything they need in order to survive when they are hatched. Mammals, on the other hand, are helpless when they are born and need to be taken care of. They have a lot to learn before they are ready to manage on their own. They learn by imitating their parents and playing with their siblings.

In order to remember what they learn, mammals have developed a new kind of memeory, the episodic memory, which is located in the hippocampus. Many cases have been reported that show that damage of the hippocampus causes loss of memory; the memories most affected are memories of events personally experienced. Because of this, it has been postulated that the hippocampus, at least initially, plays a major role in developing our individual memories.

It is precisely these memories of personal experiences that make us the individuals we are. Such memories not only represent what we saw, heard, and sensed as far as events going on around us—events that we could share with others—but of equal importance is that these memories also represent what we have felt and thought when we were faced with these events.

The hippocampus receives input from the external environment by the visual, auditory, tactile, gustatory, and olfactory senses and from the inner environment by the vagus nerve. You could say that the hippocampus plays a major role in one's sense of individuality by making a synthesis of outer and inner information.

The importance of play for memory and growth of nerve cells

When we interact with the world around us and store memories in the hippocampus, our brains change. The more signals our brains receive, the more the nerve nets and the synapses between the nerve cells are developed.

Experiments have been made with mice, showing how important external stimulation is for the development of the brain. A group of mice was raised in poor conditions, alone in simple cages, with very little sensory stimulation. Another group was raised together in a stimulating environment, with labyrinths, treadmills, ladders, and other exciting toys that were changed every day.

Psychological tests of the mice showed that those who were raised in a stimulating environment and allowed to play together were more intelligent than the other group. Their cortices were thicker, with more synapses, and the nerve cells of the hippocampus showed an increase of synapses. An important observation was that the increase of synapses could be seen only in mice who participated actively in playing and not in mice who were only allowed to look at the stimulating environment.[53]

There is no reason to believe that the human brain does not react in the same way. On the contrary, there is a lot of evidence that it does. One important conclusion you can draw from these experiments is that children must be given opportunities to stimulate and organize their brains by playing and moving around. This is especially important for infants and small children. During infancy, the nerve nets and the brain cells grow faster than during any other time. Parents cannot hope to develop the brains of their children by letting them sit in baby seats, watching what is happening around them. Even more useless is putting them in front of TVs and turn on videos about Einstein or Bach in order to make them small geniuses or wonder children. For children to develop their intelligence, they need to play and move around and be active in a joyful way. Children learn better when they play.

Stress impairs learning and memory. Stressed mice lose more cells of the hippocampus than mice that are not under stress. In stress, enkephalins are produced in the brain. These substances diminish pain, increase hyperactivity, and impair memory.

Play, imagination, and inner pictures

Unlike reptiles, mammals play. MacLean[54] discovered that the same part of the limbic system, the lobus cingulus, regulates both the mother's ability to take care of her offspring and the ability of the young ones to play. When this part of the brain was removed, the mice started to behave like reptiles: the mothers stopped caring about their young ones, and the young ones stopped playing.

The play of rodents and monkeys has many common features. It is often initiated by bouncing, followed by chasing and wrestling. These are similar features that you can see in the play of children (for instance, when playing tag).

Children also play games of make believe. Some children play more with make-believe friends than with children of their own age. Children play house or other imaginary games. In such games, they use their inner pictures and ideas. Like our emotions, our inner images are created in the limbic system, unlike the images we get from the outside world by way of the eyes.

When we read fairy tales to children, we stimulate their ability to create inner images and experience emotions, thereby stimulating their creativity. Many children have favorite tales that they ask the parents to read repeatedly. Some children love to tell stories in which they mix material from their imaginations, dreams, and fairy tales they have heard. The pictorial language of folk tales is the language of the limbic system. It is also the language of dreams, psychosis, and art. This inner language is the basis of the imagination and creativity of children and adults.

Feeling of pleasure and development of motor functions

The feelings of pleasure and discomfort are the motivation for the infant to move and experience. Feelings of pleasure are stimulated by breast-feeding and the care and gentle touch of the mother and being carried around and rocked. When the motor functions of the child have developed, play and exploration of surroundings become important for the feelings of pleasure. The child experiences distress when he hurts himself, and later, when he meets some obstacle and does not get what he wants, he will feel anger. Thus the linking up of the limbic system is closely connected with the development of motor functions.

If the infant is depressed due to neglect and lack of tactile stimulation, he will not feel like moving around, he will get less vestibular stimulation, his motor functions will not develop, and the primitive reflexes will not be integrated. In this way, the reptilian brain that is concerned with movement will not link up properly, nor will the limbic system.

The newborn baby is helpless and totally dependent on his mother. When the child develops his motor functions, he becomes less dependent on her as a source of pleasure and the means to satisfy his needs. The child learns more and more about what is fun to do and what is unpleasant or even painful.

Baby movements and the linking up of the limbic system

By approximately the age of two or three, the limbic system should have matured and the child should be ready to assert his individuality. This takes place in the assertive or defiant age, when the child learns to say "I do not want to" or "silly mother."

In order for a child to develop the nerve nets of the limbic system, he needs to develop his motor ability by making spontaneous baby movements. A child with severe motor handicaps does not go through the assertive stage, nor do autistic children. The severely handicapped child remains in symbiosis with his mother, while autistic children lose their ability of emotional attachment or never develop it. Children with fewer motor problems, who for some reason have not been able to integrate their primitive reflexes and develop their postural reflexes, will have at least some emotional problems due to a deficient function of the limbic system.

In children with motor handicaps, there is not enough stimulation of the limbic system from the proprioceptive, vestibular, and tactile sense. Such stimulation is transmitted through the reticular activation system (RAS) to the ventral tegmentum in the midbrain (the upper part of the brain stem). From the midbrain, this stimulation is conducted to different parts of the limbic system and finally to the prefrontal cortex.

The nerve nets between the limbic system and the prefrontal cortex should be sufficiently developed so that the limbic system is able to work together as a unit with the prefrontal cortex. The neurons that transmit these signals from the midbrain to the limbic system and from there to the prefrontal cortex use dopamine as transmitters of signals between each other, and therefore this system has been called the dopamine system.

Symptoms of deficient function of the limbic system

There are children who do not seem to have motor problems and yet do not go through the defiant age. They usually have retained fear paralysis and Moro reflexes, which may explain why the limbic system is not properly linked up. Some of these children may be extremely shy. Such children can develop a lifelong inability to assert themselves; when they get older, they can find it extremely difficult to say no; and they sometimes have a compulsive need to please others. They may also show a lack of curiosity about the world around them. They may disturb the play of other children and have difficulty with social interplay. These children may also show a lack of emotional responsiveness or be emotionally impulsive. They may have bouts of depression or outbursts of anger. They need not always have problems with attention and learning and sometimes may even be good at school.

Other emotional problems in children with insufficient integration of the limbic system may be, at the age of three, exaggerated reactions of defiance, with extreme temper tantrums. Some of these children have a syndrome of being extremely defiant during most of their childhoods, which may very well be carried on into adulthood. These children usually have not integrated the fear paralysis and Moro reflexes.

Motor problems in deficient function of the limbic system

As demonstrated by the case report of Tomas, symptoms of depression may occur in children whose reptilian brains are poorly linked up (e.g., with ADD or ADHD). Depressive symptoms can occur as young as the age of four and are not unusual at seven, and by the age of ten, suicidal thoughts are not uncommon in children with ADHD. Listlessness and depression are often connected with general fatigue, poor posture, feebleness, and low muscle tone. Problems with attention are also common in such cases.

Children with emotional problems, such as those mentioned above, have signs and symptoms of insufficient integration of the reptilian brain and poor arousal of the cortex. For example, they may have retained active primitive reflexes or their muscle tone may be low. Rhythmic movements, both passive and active, will stimulate the growth of the nerve nets of the limbic system and improve its function. This will improve assertiveness and power of initiative and diminish impulsiveness and fits of emotions but may also cause emotional reactions.

A motor problem often connected with emotional problems is shown in poor cooperation between the upper and lower body. Such children may not be able to move their lower bodies independently of their upper bodies and tend to involve their necks and shoulders in all their movements. Such inability is often caused by problems controlling the movements of the lumbar region.

Another cause of poor coordination between the upper and lower body may be retained reflexes. If the STNR is not integrated, movements of the head forward and backward affect the muscle tone of the lower body and make it impossible for them to move independently. People with challenges asserting themselves usually have not integrated the STNR reflex.

Reactions to the linking up of the limbic system

As stated, children with severe motor handicaps do not receive enough stimulation for the limbic system to be properly linked up. Sometimes children with less severe motor problems also may have poor function of the limbic system and the prefrontal cortex. Emotional and physical reactions and dreams that occur to people doing RMT may be a manifestation of the fact that the training causes the nerve nets of the limbic system to develop and link up more efficiently. These processes cause changes in emotions, behaviors, and even in hormone balance.

When these children first start doing RMT, it is possible they can have periods of defiance and regression. They can become demanding and babyish—hanging around their mothers and wanting to sit on their laps. They may not dare sleep alone and can dream horrible dreams and sometimes have difficulties falling asleep. In rare cases, fits of emotions will initially get worse. They want much more attention and support than the parents are used to. This is especially true for children who did not go through the assertive age. They will, after a period of RMT, start to protest and oppose, and it may be difficult for them to take part in the RMT.

These reactions correspond to crises in the emotional development of normal children. Such periods are succeeded by periods of emotional development when the children become more confident, calmer, happier, more independent, and less impulsive.

Adults can also have emotional reactions when doing RMT. Some may feel depressed and start crying without knowing why. It is also common to feel irritated or angry, and some adults have reported that they feel as if they had entered the defiant age and want to protest and oppose everything.

Causes of reactions to rhythmic movement training

Some children with a retained FPR may be oversensitive to the vestibular or proprioceptive stimulation of the rhythmic exercises, as the movements will trigger the reflex. Such children may have severe emotional reactions from the exercises. They may have temper tantrums, oppositional behavior, or be very compulsive. As will be explained in the chapter 12, the cause may be an overstimulation of the cortex and activation of glutamate receptors and accumulation of glutamate.

In many cases, intolerance to gluten and casein may worsen such symptoms. In such cases, these reactions can be avoided or mitigated by a diet free from gluten and casein.

Emotional reactions to doing RMT can also be caused by the loosening of muscle tensions that we have acquired with the long-term emotional stress and repression of our feelings of sorrow, anger, or anxiety. When these muscle armors and defensive postures start relaxing, we can react with irritation or depression as we start to release the repressed feelings.

When we get scared or angry, we contract the muscles of our legs, hips, back, shoulders, and neck, the physical aspects of the fight-or-flight response. Also, the diaphragm and the respiratory muscles of the chest are contracted. With a release of these muscles, respiration and circulation improve.

Long-term muscle tension and poor breathing has often caused the body to accumulate toxins. With the release of tension, the body can get rid of these toxins. Some of the physical reactions to the elimination of these toxins can be coughing up mucus, flatulence, nausea, diarrhea, skin rashes, itching, fever, colds, swollen eyes, headaches, fatigue, and weakness. The breaking up of physical or emotional blockages is also mirrored in our dreams. When we release muscle tensions in respiratory muscles or the diaphragm and get in touch with suppressed feelings, dreams about reptiles and mammals are common.

A mixture of the above-mentioned factors causes the emotional reactions that individuals have to RMT. A general rule of thumb is that the more emotional stress the individual has encountered during his life, and the more feelings he has repressed, the stronger the emotional reactions. Strong emotional reactions are also to be expected in individuals who, due to motor incapability, have not received sufficient vestibular and proprioceptive stimulation to link up the limbic system in a satisfactory way. Also, children with retained FPR and children with food intolerance, especially to casein and gluten, may have intense emotional reactions unless they are on an appropriate diet.

How to handle emotional reactions to RMT

When children get emotional reactions during the training, they may start to protest and not want to continue the exercises. When this happens, they may have more need for tactile and vestibular stimulation, such as massage and rocking.

In this stage, the child needs an opportunity to integrate his emotions. This stage could be considered a period of consolidation, when the brain of the child needs to take a break from the excessive stimulation that many of the movements may cause in order to integrate the new abilities and patterns of learning.

It is therefore advisable to lower the demands and exclude the movements that activate the limbic system and arouse feelings by relaxing muscle tension of the back and hips. A good option is to continue the training by doing passive rhythmic movements. Such movements stimulate both the limbic system and the prefrontal cortex, and usually the child will become more balanced and relaxed. After the child has been rhythmically stimulated for a while, it may be possible to get him to do simple movements like the windscreen wiper, sliding on the back, or rolling on the bottom. The movements will stimulate the growth of the nerve nets and the connection between the limbic and the frontal lobes, and the regressive stage will end more quickly.

The pre-birth exercises for the integration of the Moro reflex should always be discontinued since they may sometimes cause severe emotional reactions. Children with food intolerance may have excessive emotional reactions to the exercises if they are not on a specific diet. I normally recommend that such children be on a diet before they begin the rhythmic exercises.

In exceptional cases, it might be necessary to interrupt the training if the child is overly sensitive to passive rhythmic movement due to a retained fear paralysis reflex. After these children have calmed down, it is usually possible to introduce rhythmic exercises for a couple of minutes a few times a week and gradually increase the frequency and period of training.

Rhythmic movement training and dreams

Our inner images are created in the limbic system. In persons who regularly do the rhythmic exercises, such inner images are stimulated. These images usually express themselves in dreams. Only rarely can a person experience inner images while doing the exercises.

The symbolic patterns of dreams during rhythmic movement training are usually characteristic and are similar to symbols of folk tales and myths. Such dreams during RMT usually come as a confirmation of physical and psychological development processes.

Or they may throw light on unresolved psychological problems. Sometimes these dreams may be scary nightmares.

Children may have problems distinguishing what they experience in dreams from waking experience. They may also have problems telling their dreams. Sometimes children may wake up at nights and cry without being able to tell why they are afraid or sad. Obviously, they have had scary dreams.

Sometimes the contents of the dreams may be reflected in a change in how children play. Suddenly, they do not want to play with dolls or Legos, only with animals like lions or crocodiles. Dreaming about or playing with animals signifies an emotional development and linking up of the limbic system. The inner images of dreams may also become part of the fantasies and stories they like to tell.

Case Studies

Eva

Eva, whom I discussed in chapter 3, suffered from cerebral palsy and couldn't speak, eat, sit, or move around on her own when she started rhythmic movement training at the age of three. After only a few months, she started to crawl around, speak, eat, and drink. Not long after, she started to have emotional reactions. Her mother told me the following: "After we had been with Kerstin Linde for a few days, Eva started to behave so strangely that I almost got afraid. She abandoned her doll, which she never before would part from. She only wanted to play with animals. She had monkeys, tigers, and lions in her bed, and when we once entered a toy shop, she absolutely wanted a horse or a donkey, so I had to buy one for her.

"After one visit with Kerstin Linde, Eva kept smiling all the time; I think she even lay awake at night smiling. She seemed to see something, and I was afraid she had gone mad, so I phoned Kerstin. She said that Eva saw many images that she had never seen before, probably animals. Eva also started to assert herself, protest and become defiant, and say things like `silly mother´".

Olle

Olle, whom I discussed in chapter 2, also had strong emotional reactions. Before he started the rhythmic training, he was tired and sluggish and did not react when he hurt himself. He did not meet his parents' eyes and hallucinated a lot. He was not interested in his surroundings except for those times when his parents played music or sang for him, which always animated him. He was indifferent to what he ate, and his parents could stuff him with anything. He just opened his mouth and swallowed.

After a couple of months of the training, he started to have emotional reactions. He began to wake up at night and cry inconsolably. He also had a period when he was awake at night howling and laughing. He could show anger, which he had never done before, and he began to get a will of his own. He learned to protest if he did not want some food, and then he could even shake his head, which was something entirely new. And if he wanted something, he did not give up—he become very stubborn. When the family received visitors, he showed interest and crawled up to look at them. For the first time, he noticed the family dog and started to play with it. His hallucinations decreased. He began to react when he hurt himself, something he had never done before.

Both Eva and Olle are children who, because of their motor handicaps, had not been able develop the limbic system and had never learned to assert themselves. When the rhythmic exercises improved their motor functions, the limbic system started to link up and they had intense emotional reactions. A similar emotional development is to be expected in children with fewer challenges, although not so sudden.

Sam—born premature

In the following case, the situation was different. An occupational therapist who uses RMT in her work wrote me about an eight-year-old boy. He was prematurely born at 30 weeks and spent a long time in an intensive care unit. Before he started the training, he was very immature and did not function well at school, academically or socially.

After a few weeks of passive rocking from the knees and in the fetal position and actively rolling the bottom, he had enormous emotional reactions. His mother could not cope with the reactions and stopped the training. Gradually the bottom rolling exercise was reintroduced, and he could only cope with doing it one minute a week. After some time, he could do it every day, and after that the fetal rocking was reintroduced for one minute twice a week. By then, he had become a different child. He had grown in confidence and had started to assert himself, also making huge gains in schoolwork. However, he then started to lose confidence, saying he was stupid, and he became tearful at the slightest thing.

Obviously, the rhythmic exercises helped this boy to link up his limbic system and to assert himself. Also, his prefrontal cortex was linked up, enabling him not only to make huge academic progress but also to lose the innocence of childhood and get some perspective on his own limitations, which caused him to become depressed.

The triggering of his fear paralysis reflex may most likely have caused his violent emotional reactions.

Fred

Fred was nine years old when his mother brought him to me. He'd had severe sleeping problems as an infant and still had to take melatonin every night in order to sleep. He was very late to talk, and for a long time, he had problems naming things. He also had difficulties communicating with language. He also had problems socializing and playing with children of his own age. Just before his first visit, he had developed tics. He was extremely sensitive to sounds and easily disturbed. His endurance was poor, as was his balance. He had low muscle tone, and his upper arms were extremely weak, his posture shrunken. He had difficulties writing and poor handwriting. Fred had been evaluated by a psychologist and had gotten the diagnosis of attention deficit disorder (ADD).

Fred had some problems with the rhythm when sliding on his back and rolling his bottom. I tested him for dairy products and gluten with a muscle test, and he tested weak for dairy. On my recommendation, he stopped eating dairy products, and his sleep improved considerably. He did the exercises regularly every night, and during the first months, he regressed somewhat both emotionally and concerning language.

He continued to do the exercises and work with his reflexes for a year; however, there was no major improvement. His concentration and endurance were still poor, as was his ability to communicate with language. He never seemed to know what he felt, what was funny, and what was boring. He seldom protested.

After a year, during which he had conscientiously done the rhythmic exercises every day, I introduced the pre-birth exercises according to Claire Hocking for the fear paralysis reflex. He did these exercises every day and liked to do them. Now he got strong emotional reactions. He upset his teacher by protesting vehemently at school and sometimes leaving the classroom. At his next visit six weeks later, he had completely changed. He stated that he was too tired to do the exercises. He had become articulate and could describe how he was disturbed by the neighbor's motorbike. He could articulate his inner feelings. When he found school too boring, he would leave the classroom.

At his next visit after five weeks, he did not even want to see me. He preferred to go out and play in the snow, which he did for forty-five minutes. Meanwhile, his mother told me that he now refused to do the rhythmic exercises at home. However, he still did the pre-birth exercises and did not mind his mother doing the passive movements with him. He woke up at night and wanted to sleep in the parents' bed. He had started to play in a new and more focused way.

Chapter 11

The Prefrontal Cortex and Rhythmic Movements

The prefrontal cortex and the limbic system

The outer layer of the brain is the neocortex, or the human brain. Like the lower levels of the brain, its task is to process incoming signals from the senses and give adequate responses to this information. The neocortex provides the most detailed analysis and interpretation of the sensory impressions and gives the most detailed responses. There are different areas of the cortex specialized in processing different sense modalities. The occipital lobes process visual information, the temporal lobes auditory information, and the parietal lobes sensory information. The back part of the frontal lobe is in charge of voluntary motor functions. The cerebral cortex receives signals both from our inner and outer environments. The front part of the cerebral cortex, the prefrontal cortex, functions as a coordinator between the inner and outer environment and is closely connected to the limbic system. The prefrontal cortex has been defined as the areas of the frontal lobes that receive dopaminergic nerve connections from the midbrain and limbic system—that is, the signals of these nerves are transmitted by the transmitter substance dopamine. These nerve connections are called the mesocortical dopamine system.[55] The prefrontal cortex has also been called the chief executive officer (CEO) of the brain because it directs and organizes many of the processes of the brain.

The prefrontal cortex is important for many abilities (e.g., making plans and judgments, motivation, and impulse control). The prefrontal cortex enables our conceptual and abstract thinking and our ability to reason and change our conscious concepts and ideas.

The linking up of the prefrontal cortex

The prefrontal cortex has important nerve connections not only with the limbic system but also with the cerebellum, the basal ganglia, and, with the reticular activation system of the brain stem. If there is insufficient stimulation of these parts of the brain, the prefrontal cortex will not be linked up properly and the working of the whole brain will suffer. From the hippocampus, lobus cingulus, amygdala, and hypothalamus dopaminergic nerve connections lead to the prefrontal cortex.[56]

The frontal lobes of the cortex receive important stimulation from the cerebellum, going to both the prefrontal cortex and the speech area of Broca, in the left hemisphere. In cases of dysfunctions of the cerebellum, these areas may not develop properly, causing problems with speech development or difficulties with attention, ability to make good judgments, control of impulses, motivation, and sustained effort, among other things.

Also, from the basal ganglia, there are important nerve pathways to the prefrontal cortex.[57] Therefore, the working of the prefrontal cortex may be affected by motor problems, active primitive reflexes, and insufficient linking up of the basal ganglia, causing problems with attention, control of impulses, and so forth.

Retained fear paralysis and Moro reflexes may also obstruct the linking up of the prefrontal cortex. When these reflexes are triggered, the child will go into survival pattern and become aware mainly of what is going on around him. Accommodation will be obstructed, and the child will not be able to focus vision and attention and learn effectively. Instead of stimulating growth of brain cells and the linking up of the prefrontal cortex, destruction of brain cells will take place due to release of stress hormones adrenalin and cortisol.

Drive, motivation, and control of impulses

The prefrontal cortex receives nerve connections from the limbic system, bringing information about the inner conditions of the individual. In addition, it receives nerve connections from all parts of the cortex, bringing information from the senses about external conditions. When the information from the limbic system is processed in the prefrontal cortex, we become aware of our inner drives.

The amygdala and other areas of the limbic system responsible for emotions and drives like hunger, sex drive, and feelings of pleasure are closely connected with those areas of the prefrontal cortex situated on the inside of the hemispheres close to the lobus cingulus. This part of the prefrontal cortex regulates motivation and drives. Damage or insufficient linking up of this area causes shallow emotional reactions, passivity, apathy, and indifference.

In order for this area of the prefrontal cortex to function properly, it must get sufficient stimulation from the brain stem nuclei in charge of activation and arousal.

In an infant who is unable to move around due to disease, motor handicap, or low muscle tone, the prefrontal cortex is not sufficiently activated; the child may become sluggish and passive and show little or no emotions. The child may not feel any motivation to explore the world around her and remain sitting, developing neither fine nor gross motor ability.

The basal part of the prefrontal cortex plays a key role in impulse control and for the ability to stay on track and maintain a set of ongoing behavior. Damage to or insufficient linking up of this part may cause impulsivity and aggressive outbursts as well as poor foresight of the consequences of one's actions.[58]

The prefrontal cortex coordinates thought and feeling

Not until the nerve connections between the limbic system and the prefrontal cortex are sufficiently developed are we able to integrate information about external events with our inner emotions, drives, experiences, and memories and react in an appropriate way. Then we can adapt to the changes of the world around us and to our own inner needs. When, through signals from the limbic system, we become aware that we are hungry or thirsty, we can start thinking about how to get water or food. To achieve this, we must have a realistic understanding of the prevailing external conditions.

Through experience, we learn to imagine the consequences of our choices. We develop a "memory of the future," as Paul MacLean calls it. We learn which choices will bring satisfaction and which will bring pain and dissatisfaction.

The prefrontal cortex has so many connections with the limbic system that there are neuroanatomists who classify it as a part of that system. You could say that the prefrontal cortex is a comprehensive operational system, the task of which is to coordinate all our activities, talents, and creative abilities in relationship to the surroundings and to our inner needs. The prefrontal cortex is the highest level of coordination of thought and feeling.

We are dependent on the functioning of the coordination between thought and feeling. For our survival, we need to be aware of both the information we get from within, from our bodies and feelings, and information from our environment.

The linking up of the prefrontal cortex during childhood

Normally the brain of a newborn infant should be sufficiently developed for the baby to be able to feel hunger, thirst, pleasure, or discomfort. The baby should also be able to express such feelings somehow. The baby is dependent on his mother for interpreting and responding to these signals in an appropriate way.

Infants who are not aware of their internal emotional states seldom react to experiences that normally are unpleasant or pleasant. These infants have not linked up their limbic system and prefrontal cortex sufficiently and may remain passive and apathetic, without motivation to seek pleasure and explore the world around them.

When children begin school, they must be able to sit still for periods of time and concentrate on tasks that are not very exciting. They must be able to focus their attention without being distracted or diverted by external or internal stimuli. They must be able to check their impulses and function in a group, listening and doing what the teacher tells them to do. All these tasks presuppose a satisfactory linking up of the prefrontal cortex, especially the basal part of it.

Children with ADHD usually have not linked up the prefrontal cortex sufficiently by the time they start school. Therefore, they lack self-control and the ability to focus their attention. They may wander around in the classroom disturbing their classmates, not understanding that their behavior is considered thoughtless. They may have difficulties in understanding how others feel and clomp about and break conventions of social intercourse.

These children have difficulties in being able to control their impulses. They are talkative and interrupt when others talk; they have aggressive outbursts and easily get into fights. They act on impulse without reflection.

At the age of eight or nine, children make another developmental step in the linking up of the prefrontal cortex. Logical thinking becomes more prominent at the expense of the symbolic imaginative thinking. Children begin to be able to imagine the future and to compare themselves with others. They understand that their parents cannot manage everything, and they are confronted by their own limitations and lose many of their dreams. At the age of nine, many children go through a crisis and can become depressed and passive or defiant and act out.

Children with ADHD can have special difficulties during this stage. They wake up and become aware of their problems—whether they concern motor problems, learning problems, or problems in social intercourse—and come to the conclusion that something is wrong with them. They may become depressed and dejected or defiant and go into denial; often both may happen. When these children are asked if they want to improve anything, they usually answer, "Nothing is wrong with me; I don't want to be better at anything." As in the case of Tomas, you will often find that behind this attitude, these children have condemned themselves as hopeless failures.

The disposition of the limbic system to sensitization

The limbic system can be compared to an amplifier, having the ability to increase the intensity of feelings that control our behavior, so-called kindling. For instance, when we meet something threatening, the threat can be either external or internal fear. A mere trifle like an unexpected stop on the underground can, in certain circumstances, such as when there is extreme emotional or physical exhaustion, cause a strong reaction. The intensity of our feelings can be amplified until we run away, become paralyzed by fear, or start to fight. When this happens, we lose our ability to think clearly and see things in their proper perspective, which means that we switch off the ability of the prefrontal cortex to handle the external objective reality.

When we calm down, we may not be able to understand how we could react so strongly, until the next time we experience an unexpected stop on the underground. Then our reactions may be even more intense. It gets to the point that we only have to think about the unpleasant situation to start the reaction. This is called sensitization or kindling.

If the nerve nets between the prefrontal cortex and the limbic system are not sufficiently developed, or if the prefrontal cortex is not sufficiently stimulated from the cerebellum or the basal ganglia, we run a greater risk of switching off the prefrontal cortex and becoming overwhelmed by our emotions, causing fits of anger or anxiety. Small children have not developed the prefrontal cortex sufficiently and therefore may have violent outbursts of anger that may even end in convulsive fits. Children with ADHD and autism may also have violent outbursts of anger because of poor function of the prefrontal cortex. Children with retained Moro and fear paralysis reflexes will shut off the prefrontal cortex in stress and go into survival patterns, and they can react with either paralysis or temper tantrums.

Harald Blomberg, MD

The effect of rhythmic movement training on the prefrontal cortex

Rhythmic movement training improves the function of the prefrontal cortex by stimulating the reticular activation system, the limbic system, the basal ganglia, and most importantly, the cerebellum—and by developing the nerve nets of these areas so that the prefrontal cortex may continue to receive enough stimulation for its proper functioning.

In children with a retained fear paralysis reflex, the rhythmic movements may sometimes cause severe emotional reactions because the activated reflex causes kindling of the limbic system. Normally this does not happen because the simultaneous stimulation of the prefrontal cortex by way of the cerebellum and possibly the basal ganglia will moderate and balance the tendency of kindling of the limbic system. Usually one of the first signs of improvement during rhythmic training, especially in ADHD, is a decrease in impulsivity and emotional fits.

Children with ADHD who do the rhythmic exercises will often notice improvement in attention and concentration within a couple of months. They learn to stay on track without being distracted. Their ability to reason logically usually improves, and they may notice that mathematics is not as difficult as it used to be. Also, their self-confidence and self-esteem improve; they become happier and more outgoing, and the contact with peers improves.

In some cases, this development takes more time. For instance, this may be seen in some children who have great difficulties learning how to make the rhythmic movements in a smooth and coordinated way. Consequently, their prefrontal cortices will get less stimulation from the cerebellum than in children who have no problem with the rhythm of the movements.

Problems with the rhythmic exercises reflect a malfunction of the cerebellum, which may be caused by inflammation, usually due to intolerance to casein or gluten. Such intolerance may cause excessive emotional reactions to the rhythmic exercises, and if this happens, a diet is important to decrease the inflammation of the cerebellum and improve the stimulation of the prefrontal cortex.

Chapter 12

Autism Spectrum Disorder and Rhythmic Movements

What is autism and how did the diagnosis come into existence?

In 1943, American child psychiatrist Leo Kanner described a new syndrome in children in a textbook with the title *Autistic Disturbances of Affective Contact*. Some of the symptoms of the children he described were inability to relate themselves to people, poor or absent language skills, sensory sensitivity, repetitive behavior, and an obsessive desire for maintenance of sameness.

In 1980, the diagnosis "infantile autism" was included in the DSM III, the American diagnostic manual for psychiatric diseases, with the following criteria: (1) onset before thirty months of age; (2) pervasive lack of response to other people; and (3) gross deficits in language development or peculiar speech patterns.

This definition was considered too restrictive and was modified in 1987. Now the criteria became abnormal social interaction, abnormal communication, and narrowed interests or activities. This was exemplified with sixteen separate criteria, of which a certain number were required for the diagnosis.

In 1994, the criteria for autism were modified again in order to narrow the diagnosis. Related disorders that did not fulfill all the criteria of autism (e.g., Asperger's syndrome) were included in the diagnosis autism spectrum disorder.[59]

Dramatic increase in children diagnosed with autism

Before 1980, autism was a rare disorder. Scientific studies consistently estimated the rate to be 2 to 5 per 10,000 people.

In the following twenty-five years, there was a rapid increase in the frequency of autism. In 2004, an official report from the Centers for Disease Control (CDC) established that the current incidence of children with autism and autism spectrum disorders (ASD) in the United States was 1 in 166.

In 2012, the Centers for Disease Control (CDC) announced that 1 in 88 children in the United States was diagnosed with an autism spectrum disorder (ASD). The updated estimates are based on data collected in fourteen American communities during 2008.

This development can be illustrated with what happened in California. In California, there are regional centers to care for people with disabilities. About 75 to 80 percent of developmentally disabled children were enrolled in the system.

Before 1980, the number of enrolled children with autism was between 150 and 200. In 1987, this number had risen to 400. During the following ten years, there was a rapid increase. Between 1987 and 1998, there was an almost threefold increase of children diagnosed with autism. The number of enrolled children with ASD had increased almost twenty times. In 2000, more than 20,000 children with ASD were enrolled in the regional centers, as compared with 200 children with autism in 1980.[60]

Other studies have confirmed a similar increase in cases of autism in other parts of the United States. The increase in Europe has been substantial but not as rapid as in the States.

What caused the increased incidence of autism?

According to academic medicine, autism is a chronic neurological disability with hereditary genetic causes. However, an increasing amount of research during the last twenty years shows that children develop autism because their the immune systems and detoxification abilities no longer can cope with increasing environmental stress from vaccinations, heavy metals, and other toxic substances.

Academic medicine has made great efforts to deny the increasing incidence of autism since an epidemic of a hereditary genetic disorder is considered to be impossible. However, a recent study has shown that autism may be caused by damaged genes that have not been inherited from the parents.[61] This discovery confirms the importance of the environment as a causal factor in autism.

Underlying causes: mercury and vaccinations

Children who develop autism have an impaired ability to cleanse the body from mercury and other heavy metals that accumulate in the brain and inner organs. Mercury is one of the most toxic substances that exists and is especially harmful to the neural system and to fetuses. One important source of mercury is amalgam, which releases mercury into the body. There is evidence that pregnant mothers pass on mercury from amalgam to a fetus, which will accumulate it in the developing neural system. A recent study of one hundred women has shown that if a mother has more than eight amalgam fillings, her child runs a more than four times increased risk of developing autism than children of mothers without amalgam.[62]

Another source of mercury is vaccinations. In the beginning of the 1990s, the number of vaccinations given to children containing thimerosal increased in a drastic way in the United States. Thimerosal is a preserving agent widely used in vaccines and containing mercury. According to a 2005 article in the *Boston Globe*[63] by Robert Kennedy Jr., one of the fathers of Merck's vaccination programs, Dr. Maurice Hilleman, in a 1991 memo, warned his bosses that six-month-old children administered the shots on schedule would suffer mercury exposures eighty-seven times the government safety standards. He recommended that thimerosal be discontinued and complained that the US Food and Drug Administration, which has a notoriously close relationship with the pharmaceutical industry, could not be counted on to take appropriate action as its European counterparts had.

Dr. Hilleman's warning went unheeded, and the vaccination program was allowed to continue, causing an enormous epidemic of autism. In ten years, the autism rate in American children increased from 1 in 2,500 children in 1995 to 1 in 166 children or 1 in 80 boys in 2005. In addition, one child in six was diagnosed with a related neurological disorder.

The harmful effects of thimerosal

Numerous scientific studies have been conducted that point to thimerosal as a culprit in this epidemic. Autistic children have been shown to have higher mercury loads than children without autism. Many symptoms of autism are similar to the symptoms of mercury toxicity. A study by an FDA scientist, Dr. Jill James, found that many autistic children are genetically deficient in their capacity to produce glutathione, an antioxidant generated in the brain that helps remove mercury.

Drug makers ceased to use thimerosal to produce vaccines after 2000 to avoid liability, but vaccines stocks with thimerosal were not destroyed and continued to be sold until 2004. Health authorities are still defending the use of thimerosal in vaccines. Robert Kennedy Jr. concludes his article in this way: "Government officials who continue to champion thimerosal should recognize ... that they are depriving vulnerable populations from being identified to avoid thimerosal. They also cannot escape responsibility for their failure to warn international health agencies and governments who, based upon American assurances, are now injecting the developing world's children with this brain-killing chemical."

The MMR vaccination

Another devastating effect of vaccinations is damage to the immune system of the intestines caused by the MMR vaccination (measles, mumps, and rubella). Statistics from California show a significant increase of autism after the introduction of the MMR vaccine in

1978. Three years after Californian children started to receive this vaccination, the number of children diagnosed with autism trebled.[64]

British scientist Andrew Wakefield demonstrated a connection between MMR vaccination and autism. Such vaccinations can cause swelling of the lymph nodes in the bowels and impair the assimilation of nourishment; the swelling can also cause constipation. Wakefield and his colleagues have been able to isolate the same measles virus as in the MMR vaccine from the intestine of autistic children with bowel symptoms.[65] Since Wakefield published his first article, a number of scientific articles have been published that corroborate his results. Several studies have confirmed the presence of the same type of measles virus in lymph nodes from the bowels and in the blood of autistic children as can be found in the MMR vaccine.[66]

In the UK, more than two thousand families claim that their children developed normally until they were given the MMR vaccination at around the age of twelve or eighteen months. In many cases, they came down with a fever after the vaccination and suddenly regressed, stopped talking and playing as before, and developed typical autistic symptoms. The British Department of Health and the British Medical Association claim that the vaccine is safe, citing numerous statistical studies. However, when I have heard so many parents relate how their children became autistic immediately after the MMR vaccination, I am convinced that this is just another instance when statistics are used not to reveal the truth but to conceal.

Intolerance to gluten and casein in autism

In the beginning of the 1990s, aDr. Mårten Kalling told me about a Norwegian doctor and researcher named Karl Reichelt, whom he had met at a medical conference, at which Dr. Reichelt presented his research about treating autism and schizophrenia with a gluten- and dairy-free diet. According to his theory, an impaired functioning of the bowels prevents the breaking down of casein (milk protein) and gluten. Instead of amino acids, peptides are formed, called exorphines, and these have an effect similar to morphine. About half the children with autism had been found to secrete such peptide chains in their urine.[67]

Dr. Reichelt has described two different patterns of peptides in the urine of autistic children. One of these patterns is typical of children who are intolerant to casein and the other of children who are intolerant to gluten. The second pattern can be found in children who got autistic symptoms later, after a normal development during the first one or two years, called regressive autism. Parents of these children often reported that the children developed digestive symptoms and continuous crying after weaning, when cereals and cow's milk were introduced in their diet, and sometimes after vaccination, usually flu vaccination or the MMR vaccination.

In autistic children, these peptides affect play and social interaction and may cause self-mutilating behavior and reduced sensitivity to pain. These exorphines are highly addictive and often cause children to limit their diets to dairy products and wheat.

In autism, another harmful effect of gluten and casein is the production of cytokines, which are substances that cause inflammatory reactions, especially in the cerebellum.

Celiac disease and gluten sensitivity in autism

In 2012, leading scientists who have done extensive research on gluten have reached consensus on a new classification of gluten-related disorders.[68] They differentiated between celiac disease, which can be diagnosed by blood tests, and intestinal biopsy and gluten sensitivity. The latter condition cannot be diagnosed by any tests, only by observing how clinical symptoms improve by a gluten-free diet. The two conditions cannot be distinguished clinically since their symptoms are similar. It is believed that gluten sensitivity is at least six times more widespread than celiac disease.

A minority of children on the autistic spectrum are diagnosed with celiac disease. Celiac disease has been shown to be three and a half times more common in patients on the autistic spectrum than in normal individuals. This would imply that at least one child out of thirty with autism could suffer from celiac disease.[69]

The majority of children with autism are gluten sensitive (i.e., they regress when they eat gluten), and there are no biomarkers for celiac disease in their blood. The only way to diagnose this condition is to study how health and well-being improve by eliminating gluten. The regression due to gluten in autism can be explained by the production of gluteomorphines in the gut, which cross the intestinal wall and enter the brain, producing typical symptoms because of both morphine effects and inflammatory and immune reactions.

I regularly have recommended a diet free from gluten, casein, and soy to patients with psychosis and autism and have seen remarkable improvements in the majority of cases. Also, many parents of children with autism have tried the gluten- and casein-free diets for their children and have corroborated the beneficial effects on their behavior, provided the children were able to adhere to a strict diet.

Two controlled randomized clinical studies of diet intervention in autism have been published. In a Norwegian study from 2002, twenty autistic children underwent developmental and behavioral testing. Half of the children were put on a gluten- and casein-free diet for one year, after which the tests were repeated. The children in the diet group showed significant improvement, while there was no change in the control group.[70]

In a ScanBrit study, seventy-two Danish autistic children from four to ten years of age were tested and randomly assigned to a diet or a non-diet group. After six and twelve months, the tests were repeated, and the results showed clinical improvement in the children on a diet compared to the other group. Their attention improved, and they became less aggressive, less hyperactive, and started to talk more. Many of the children got rid of their allergies, and in several of them, epileptic seizures diminished. The conclusion was that the diet helps two-thirds of the children.[71]

Case study: Carl

I especially remember a seventeen-year-old autistic boy named Carl. He was a student of an anthroposophic boarding school. He had deteriorated severely during the previous months and had stopped talking. He did not go to classes, he refused to go for meals, and he spent his days in his bed, not communicating and hurting himself severely by pounding his face with his fists. I was called on to do RMT with him. When I asked what he used to eat, I was told that he survived on milk and biscuits, which he took from the kitchen at night. He did not eat anything else. I told the staff to exclude all gluten and dairy from his food before I was willing to start rhythmic training.

After a couple of months, I was asked to give a lecture on RMT at the school. The staff informed me that Carl had recovered dramatically after they had managed to change his diet. He now went to school and had stopped mutilating himself; he talked and communicated as he had before he deteriorated.

Carl continued his diet as long as he stayed at the school and got on fine. Then he moved to a group home. They needed a doctor's certificate that he was intolerant to gluten and dairy products, which I was asked to write. However, before I sent it, I was informed that they did not need it anymore. They had given him bread and milk from the very first day, and now he was consuming them without a problem. I was very surprised and asked if his health had been affected after he moved to the group home. I was told that he got psychotic within a week and had to be admitted to the hospital, where he was heavily medicated with neuroleptics. He was still on medication.

Gluten, dairy products, and psychosis

This episode reminded me about what Dr. Reichelt had told me when I met him. He believes that neuroleptics not only affect the brain but also somehow the intestines, and that they might decrease the assimilation of the peptides, which may cause psychosis.

During my work as a school doctor at a Waldorf boarding school, I have repeatedly learned that becoming psychotic is not an occasional incident but rather what may normally happen at a certain age to a certain group of students. Young men diagnosed with autism or Asperger's, especially those who also have bowel problems like constipation or loose stools, run the risk of becoming psychotic around the age of eighteen to twenty, sometimes before. In many of these cases, I have established that they do not tolerate gluten and casein, and although some of them have been able to exclude the latter from their diets, their motivation had not been sufficient to exclude gluten. However, experience has taught me that provided they get a moderate dose of neuroleptics, these patients can go on eating gluten and in many cases do not relapse into psychosis. And in some severe cases of constipation, it seems as if the bowel problems also improve by taking the medication.

EMFs and microwaves

To my knowledge, no studies have been made about electromagnetic fields from mobile phones, cordless phones, or cordless networks causing autism. But there is extensive evidence that such radiation may cause the very kind of damage that is common in autism. According to the big REFLEX study commissioned by the European Union, such radiation is the cause of free radicals that may damage the DNA and the cell membranes and cause cell death.

In 2007, a group of fourteen scientists published the Bioinitiative Report, which was based on fifteen hundred studies of the effects of electromagnetic irradiation. They found that genes are damaged even by radiation below accepted safety standards, and they found extensive evidence that electromagnetic fields cause inflammation and allergies and weaken the immune system.[72]

Considering the fact that most fathers-to-be carry their mobile phones in their pants pockets, the findings of damaged genes causing autism are easily explained. The fetuses, whose genes are already damaged before conception, will then be exposed to even more radiation from cordless networks and their mothers talking on their mobile phones or cordless phones. Knowing that all this radiation may damage cell membranes, kill brain cells, damage the immune system, and cause inflammation of the brain and the bowels, it is easy to explain why many children are so weakened when they are born that they develop autism, especially when they are exposed to additional environmental stress such as vaccinations, virus infections, and heavy metals, to say nothing of continuing microwave irradiation.

When the cell is exposed to electromagnetic irradiation, it produces stress proteins that make the cell membrane less penetrable and reduce the active transport of toxins and heavy metals over the membrane. Elimination of electromagnetic irradiation is therefore necessary to give the child a basis to detoxify the cells from heavy metals and other toxic substances.

The glutamate to GABA ratio

Glutamate is the most common transmitter substance of the synapses of the brain. It has a stimulating, excitatory effect and causes cellular toxicity if too much accumulates. Glutamate is converted into GABA, the function of which is to inhibit the firing of the nerve cells. Several studies have shown an imbalance between glutamate and GABA in autism due to deficient ability to convert glutamate into GABA.[73]

An accumulation of glutamate and a lack of GABA will trigger uninhibited firing of the nerve cells, which will cause inflammation and eventually cell death. When the level of glutamate increases, GABA will diminish, which may cause the child to stop talking.

When the brain of an autistic child gets too much stimulation from the visual, auditory, tactile, or vestibular senses—whether from physical activity, psychological stress, or peptides from the bowels—the accumulation of glutamate may cause the child to become overstimulated and hyperactive and have great difficulty calming down. Because of accumulation of glutamate, the neurons of the brain continue firing, causing self-stimulating behavior and inflammation of the brain and in the worst case brain damage.

Such reactions are considered to cause similar brain damage in Parkinson's, MS, and ALS as in autism. Inflammation of the brain due to accumulation of glutamate has also been linked to epileptic seizures, which are common in autism.

Self-stimulating behavior and symptoms of inflammation of the brain

Self-stimulating behavior, or "stims," indicate an accumulation of glutamate in the brain. Common stims include hand flapping, body spinning and rocking, lining up or spinning toys or other objects, perseveration, and repeating rote phrases.[74]

Uncontrolled firing of neurons due to elevated levels of glutamate causes inflammation and injury of the nerve cells. As stated, peptides from the bowels may also cause inflammation of the brain. Symptoms of such inflammation may be insomnia, epileptic seizures, obsessive-compulsive behaviors, aggressiveness, temper tantrums, depression, anxiety, and psychosis.[75]

Rhythmic movement training in autism

Passive rhythmic exercises usually work well for children with autism, causing them to become calm and relaxed. However, if the brain becomes overstimulated by the exercises, they will become restless and anxious. Such reactions are sometimes seen also in children with ADHD and ADD. Both children with autism and ADHD may have a reduced ability to filter sensory impression in the brain stem due to a retained fear paralysis reflex, causing overstimulation of the cortex and activation of glutamate receptors and accumulation of glutamate. This may cause the child to squirm and struggle to get away. Autistic children may also react with self-stimulating behavior.

If this happens, you need to start very gently with the exercises and stop immediately when the child starts to wriggle. You may be able to repeat them just a few times and then gradually increase them. In this way, the child will learn to put up with the exercises for longer periods without negative side effects, possibly because the exercises improve the filtering ability of the brain stem or because the passive exercises stimulate glutamate to convert into GABA. It may also be effective to rock the child gently when he is asleep.

Rhythmic movement training, autism, and seizures

Epilepsy is not uncommon in autism. Seizures are a symptom of inflammation of the brain that may be caused by excessive levels of glutamate. As we have seen, an accumulation of glutamate and inflammation of the brain may be caused by intolerance to gluten and dairy. According to Reichelt, a diet free from gluten and dairy will reduce seizures. Other causes of accumulation of glutamate in the brain are food additives like monosodium glutamate (MSG) and aspartame, which are common in processed food. Both MSG and aspartame are notorious for causing seizures. Seizures may also be triggered by electromagnetic radiation.

In order to avoid triggering seizures when doing rhythmic exercises, it is essential to prevent an accumulation of glutamate. Therefore, a diet free from dairy and gluten, MSG, and aspartame is advisable in autistic children and of course in all children with known epilepsy. It is also very important to avoid self-stimulating behavior by doing the exercises very gently in the beginning and gradually increasing them. Elimination of electromagnetic irradiation, especially in the child's bedroom, is also of great importance in order to prevent seizures.

Rhythmic exercises and the cerebellum

Many children with autism are unable to do simple active rhythmic exercises when they start RMT and do them in a characteristic way. They cannot do them automatically. I have found that they have to use a lot of effort for every turn to the right or left or when they roll the bottom from side to side. Unlike other children with problems of the cerebellum, they do not lose the rhythm when they do these exercises. They simply have no rhythm at all. This inability is caused by a dysfunction or damage of the cerebellum.

Many studies have shown that damage of the cerebellum due to inflammation or toxic processes is very common in autism; especially the Purkinje cells that use GABA as their transmitter substance are affected. Moreover, the cerebellum in autism is often smaller than normal, as is its nucleus dentatus, which is linked to the speech areas of the left hemisphere by an important nerve path. When this nucleus is damaged, language will be affected. Consequently, lack of speech, deficient speech, or late speech development is common in autism.

RMT normally has a good effect on language development in autism. The active exercises stimulate the Purkinje cells and GABA. The exercises indirectly stimulate the speech areas of the left hemisphere. The exercises therefore improve both language comprehension and spoken language, provided the harmful inflammation process of the cerebellum has been brought under control with an appropriate diet. In some children, the cortical speech areas of the left hemisphere are damaged as well. In such cases, language will be slower to develop.

Simple rhythmic exercises such as the windscreen wipers or rolling the bottom are also excellent to stimulate the prefrontal cortex by way of the cerebellum. When children learn to do these exercises in a rhythmic way, the cerebellum will be able to stimulate the function of the prefrontal cortex, thus improving impulse control, endurance, judgment and empathy.

Rhythmic movements and the limbic system in autism

The passive and the active rhythmic exercises stimulate the limbic system by way of the ventral tegmentum in the midbrain in the upper part of the brain stem. When the limbic system is stimulated by the exercises, emotions will develop. The children usually will start to show emotional attachment to their parents, and for the first time ever, they may want to sit on their laps or be hugged. Compulsive symptoms and ritualistic behavior may diminish, and they usually begin to show emotions of defiance and self-assertion. They often start to play in a different way, as well as also starting to play with other children, and many parents report that they start to express empathy.

Some children, especially after puberty, can have problems in handling feelings of assertion, defiance, and aggression that may occur during RMT. Sometimes aggressiveness and temper outbursts may increase, especially in mentally retarded children. Such reactions are usually caused by excessive glutamate levels and aggravated by intolerance to gluten and casein in the way that has been explained. These reactions can be avoided or mitigated by a diet free from gluten and casein and elimination of electromagnetic irradiation.

Retained stress reflexes in autism

Most children with autism or ASD have oversensitive senses and are very sensitive to stress due to retained fear paralysis and Moro reflexes. An active fear paralysis reflex often causes vestibular sensitivity and may cause the child to become nauseated during passive exercises. The vestibular, tactile, and proprioceptive sensitivity may also contribute to stims when the brain is overstimulated by the exercises. It is therefore very important to integrate this reflex in order to be able to continue the movement training.

As stated, the fear paralysis reflex starts to develop in the second month after conception. At that age, the fetus is busy moving his recently developed arms and legs most of the time. When the fetus is exposed to stress, the fear paralysis reflex is triggered (i.e., the fetus is paralyzed and stops moving).

Again, the reflex is not a primitive reflex, since neither the senses nor the neural system has been sufficiently developed to generate any reflex pattern. Instead, the reflex pattern is created by direct transmission of information between the cells instantly to all cells of the fetus by electromagnetic frequencies.

The fear paralysis reflex is triggered when the cells of the fetus are exposed to stress. They shut off the environment by producing stress proteins that make the cell membrane hard to penetrate and diminish

The Rhythmic Movement Method

the active transport over the membrane. When this happens, the fetus stops moving and becomes paralyzed.

As previously stated, different kinds of stress may prevent the reflex from integrating. Stress in the mother due to depression, illness, or drug abuse as well as stress from the environment (e.g., mercury or other heavy metals, toxic substances, or electromagnetic fields) are such factors.

When the fear paralysis reflex is not integrated, the Moro reflex will not integrate. In most cases, it is therefore advisable to integrate the fear paralysis reflex before starting to integrate the Moro reflex. Even if the fear paralysis reflex was integrated in the womb, it can be activated at any time by excessive levels of environmental stress, such as electromagnetic radiation or intolerance to gluten and casein.

Many symptoms of autism are also symptoms of a retained fear paralysis reflex

Many symptoms of a retained fear paralysis reflex are also common symptoms of autism, including low stress tolerance and oversensitivity to sensory stimulation such as sound, light, smell, and taste. The reflex will be triggered by eye contact, and poor eye contact is a common symptom both of autism and a retained fear paralysis reflex. When the reflex is retained, the child needs to withdraw from challenging stimuli and situations that are perceived as threatening. Children with autism may shut off the world around them and stop communicating. Some children may become very shy. Other strategies to avoid challenging situations may be sticking to routines, poor adaptability, and refusal to changes, all common symptoms in autism. When the child is not able to shut out the world, severe temper tantrums are common reactions.

Integration of the fear paralysis reflex

Since the fear paralysis reflex is a cell reaction to stress, the cause of the stress must be dealt with in order to integrate the reflex. If the reflex is constantly triggered by electromagnetic irradiation, it cannot be integrated until the electromagnetic stress has been reduced. Likewise, if the reflex is caused by toxicity in the environment or the body or food intolerance, these causes must also be taken care of. In autism, you need to reduce electromagnetic stress, heal the bowels by having a proper diet, cleanse the body from heavy metals, and so forth. In ADHD and ADD, you may also have to deal with such issues, especially electromagnetic stress and intolerance to casein and gluten.

Active and passive rhythmic movements also contribute to integrating the reflex, as do the pre-birth exercises as taught by Claire Hocking. The latter movements are similar to the movements the fetus makes in the womb. These exercises are sometimes very powerful and may trigger the reflex and cause negative reactions. That seems to be especially true if there is a lot of environmental stress. In such cases, this stress should be taken care of first. In my experience, these exercises usually fail to integrate the reflex if there is gluten and dairy sensitivity, unless the child is on a proper diet.

CHAPTER 13

RHYTHMIC MOVEMENT TRAINING AND PSYCHOSIS

Increased activity in the limbic system in psychosis

As stated, the limbic system could be compared to an amplifier, being able to increase or decrease the intensity of the feelings that control our behavior.

If a person is allowed to float in a solution of salt in a completely dark and silent room, after some time, the person will react with intense anxiety and visual and auditory hallucinations. In this condition of so-called sensory deprivation, the cortex of the brain gets no stimulation. There is no awareness of outer stimuli, and the mind will instead focus on inner processes of the limbic system. When the activity of the prefrontal cortex decreases, it will increase in the limbic system due to kindling, which explains the symptoms.

A Swedish study has shown that people diagnosed with schizophrenia have decreased activity of the prefrontal cortex, especially on the left side. At rest, the blood flow of the brain of healthy individuals is mainly concentrated to the frontal lobes, especially the prefrontal cortex. However, in persons diagnosed with schizophrenia, the blood flow of these areas is conspicuously absent. Instead, the blood flow is increased in the parts of the temporal lobes that belong to the limbic system.[76]

Because of this pattern of activity, the prefrontal cortex will not be able to regulate the activity of the limbic system. When there is a strong emotional stimulation (e.g., falling in love, the death of a close relative, or what have you), such a person may lose contact with reality and experience more or less frightening inner symbolic images as outer reality. In even more extreme cases, a catatonic condition may develop when the person loses the ability to react to external stimuli and the consciousness is totally focused on inner images (hallucinations). This condition is related to the condition of Olle before he started rhythmic exercises, when he was hallucinating and did not react to external stimuli. As we have seen, Olle suffered from a severe lack of stimulation of his brain from his senses and as a result a decreased activity of his prefrontal cortex.

When a person no longer reacts to external stimuli, she has developed a condition of total inner sensory deprivation. In spite of normal outer stimulation, the prefrontal cortex is not sufficiently activated in order to maintain its activity.

Intolerance of gluten and casein in schizophrenia

As stated in the previous chapter, the limbic system may also be triggered by peptides that arise in the bowels due to intolerance to casein and gluten in combination with a lack of peptidase. Reichelt has shown that a diet without gluten and casein will improve symptoms in schizophrenia.[77]

The connection between gluten intolerance is not new knowledge. It has been recognized since the 1960s, when an American psychiatrist named F. C. Dohan started to research this subject. He had learned that gluten intolerance was much more common in schizophrenic patients than would be statistically expected. He and a collaborator studied schizophrenic patients that were given a gluten- and milk-free diet after being admitted to the hospital due to a deterioration of their disease. They could show that the patients who received the diet recovered more quickly and could be discharged in half the time compared to a control group without this diet. Additional studies showed that patients who recently had fallen ill benefitted most from this kind of diet, while chronic patients who had been ill for many years only improved in exceptional cases.[78]

Young people who have recently been treated for psychosis and are looking for an alternative to medication sometimes consult me. They or their parents may have heard or read about RMT. I normally recommend that they exclude gluten and dairy products from their diet, especially if I have established by bio- resonance that they do not tolerate these foods. I explain that they will benefit more from the rhythmic movement training if they change their diet, not only because gluten and milk products may cause a psychosis by themselves but also because they may trigger strong emotions if not tolerated. Such emotions may trigger a psychosis if the function of the prefrontal cortex is insufficient. The rhythmic exercises will stimulate the prefrontal cortex to develop but may also trigger strong emotions. Additionally, gluten and casein may cause inflammation of the brain, which will worsen the disorder in the long run.

I therefore discourage them from stopping medication while they are doing the rhythmic movements, at least in the beginning, before the nerve nets of the prefrontal cortex have had a chance to develop more efficiently.

Negative overshooting feedback

It is possible to understand the cause of inner sensory deprivation in psychosis by looking at the role of the brain stem in maintaining the activity of the limbic system and the prefrontal cortex.

Dopamine-producing nerve cell nuclei of the midbrain part of the brain stem send their nerve fibers in nerve paths to the limbic system and prefrontal cortex. The function of the mesocortical dopamine system is, among other things, to regulate the activity of the limbic system. This regulating activity could be compared to the mode of operation of a ventilation system of a house. The function of such a ventilation system is to keep an even and stable temperature in all rooms. With feedback, the effect of the heating station will increase or decrease according to need. When the temperature of one or several rooms rises, the effect of the heating station will decrease.[79]

If this model is used in schizophrenia, one can understand how an activation of the limbic system combined with a disorder of the prefrontal cortex may trigger the psychosis. When the limbic system is kindled by psychotic delusions, hallucinations, and intense emotions, the compensatory regulation will be activated. With feedback from the limbic system, the nerve nuclei of the midbrain will decrease their activating signals to the limbic system and the prefrontal cortex. However, due to the kindling, the activity of the limbic system will not be affected. Only the activity of prefrontal cortex will diminish, creating a vicious circle that may eventually lead to a catatonic state; the more the limbic system is kindled, the less the activity will be of the prefrontal cortex. The person will lose her basis in reality and ability of rational thinking and will become prey to distressing emotions.

Rhythmic movement training in acute psychosis: a case study

The hyperactivity of the limbic system in psychosis can be treated with drugs that block the function of dopamine, called neuroleptics. Such treatment will diminish hallucinations and delusions after a few days. However, neuroleptics will also block the activity of the prefrontal cortex, causing side effects such as mental and emotional blunting.

A different strategy of treatment would be to increase the activity of the prefrontal cortex in order to create a balance between the limbic system and the prefrontal cortex. This can be done by rhythmic exercises, and in cases of acute catatonia, you do not have to wait several days for the treatment to have effect.

One of my patients had developed increasing anxiety and delusions after an emotionally traumatic experience. Not wanting to take neuroleptics due to the side effects, she started to do rhythmic exercises every day. Her condition improved during the first months, but then she deteriorated and stopped doing the exercises. She became bedridden and catatonic and lost her ability to communicate with her family. After a couple of days in this condition, she was brought to the hospital.

I met her in the ward soon after she was admitted and before she had received any medication. I could not get through to her, and I started to do passive rhythmic exercises with her. After a few minutes, she was able to do the rhythmic exercises on her own, and after about fifteen minutes of rhythmic exercises, she could sit up and talk to me freely. She told me that during the last two days, she had had intense and frightening hallucinations about war and her house being bombed. I continued to help her to do rhythmic exercises in the ward every day. She was also prescribed a very small dose of neuroleptics. She recovered rapidly and could be discharged after a couple of weeks. She continued the rhythmic exercises and could soon discontinue the medication, and after a year, she had recovered completely and could start working again. She then worked for another fifteen years without problems before she retired.

The mode of action of RMT in psychosis

The rhythmic exercises have a strong arousing effect on the cortex and especially the prefrontal cortex by way of the reticular activation system (RAS). But this is not the only reason that the rhythmic exercises have an immediate effect on delusions and hallucinations, as has been described. The rhythmic exercises also affect the prefrontal cortex by way of the cerebellum, thus shortcutting the limbic system.

Research has demonstrated that the cerebellum plays an important role in increasing the activity of the prefrontal cortex and decreasing psychotic symptoms like hallucinations. In 1977, an American scientist named R. G. Heath published an article about treating a group of severely ill schizophrenic patients who had been hospitalized for many years and had not responded to any treatment.[80] He inserted a pacemaker into the cerebellum of each person of this group. As previously stated, the cerebellum is closely connected with the prefrontal by important nerve paths. The patients of the study could regulate the activity of the pacemaker. When they started to hallucinate, they only needed to turn the knob to increase the stimulation and the hallucinations would subside. Most of the patients improved conspicuously and several could be discharged. Since the rhythmic exercises stimulated the cerebellum greatly, it can be concluded that they have a similar mode of action as the pacemaker (i.e., stimulating the prefrontal cortex by way of the cerebellum).

In another study, Heath and his coworkers could show that a high percentage of people who had been diagnosed with schizophrenia (between 33 and 60 percent) suffered from atrophy (decrease of cells) of the area of the cerebellum (the vermis), which has important pathways to the prefrontal cortex.[81] Impaired function of the vermis could explain why the prefrontal cortex gets insufficient stimulation. A common cause of atrophy of the vermis is heavy metal toxicity.

The atrophy of the vermis and the impaired function of the prefrontal cortex in schizophrenia can explain how the inner sensory deprivation can be so conspicuous in some cases of schizophrenia.

Positive and negative symptoms in schizophrenia

Psychotic symptoms like hallucinations and delusions are called positive symptoms and are caused by negative overshooting feedback triggering a hyperactivity of the limbic system and a shutting down of the prefrontal cortex. The impaired function of the prefrontal cortex in schizophrenia also causes negative symptoms e.g. emotional blunting, passivity, indifference toward those around and poor judgment. Such negative symptoms are common among schizophrenic patients who have been ill for a long time.

Rhythmic movement training may improve the symptoms of psychosis in different ways. In acute cases with severe positive symptoms the hallucinations and delusions may decrease because the rhythmic exercises stimulate the activity of the prefrontal cortex creating a balance between the limbic system and the prefrontal cortex. In chronic schizophrenia with severe negative symptoms the stimulation of the prefrontal cortex will usually improve the negative symptoms and cause the patients to become more interested in people around, less withdrawn, more active etc.

Psychosis as a language disorder

An important factor that contributes to the long-term recovery of psychosis is the fact that anxiety decreases when frightening psychotic delusions are taking shape in dreams during rhythmic movement training. By way of the dreams the patient learns to discriminate between inner symbols and outer reality, which helps to reduce anxiety. When the anxiety decreases the limbic system will not be activated in an uncontrolled way making it possible to keep inner balance and harmony and avoid psychosis.

A psychotic person has more or less lost his ability to discriminate between inner and outer reality and often perceives his inner symbolic world as outer reality. Such inner symbols could be a manifestation of inner blockages or malfunctions on a physical or emotional level. In psychosis such inner symbols come to the forefront. Kerstin Linde used to say that psychosis is a condition when the person has turned his eyes inward and uses symbolic language to explain his problems to himself and to those around.

Rhythmic movement training allows physical and emotional blockages to take shape in dream symbols. This will permit a dialogue between the language of the cerebral cortex, i.e. spoken language that has been learned, and the inborn symbolic language of the limbic system. When a psychotic person dreams a dream in which her delusions are taking shape she may understand that the delusions she has been fostering are not outer reality but only inner dream symbols. This may come as a great relief since the experienced reality in psychosis may be so much more frightening than the dream experience that we wake up from. By such a dialogue between the symbolic language of the limbic system and the spoken language of the cerebral cortex the psychosis may heal.

Case study: Lotta

The case of a patient I treated in the 1980s may illustrate the healing effects of dreams during rhythmic movement training. Lotta was a woman around thirty who had been suffering from severe delusions over a couple of years. She was afraid to be murdered and in periods she did not dare to go out. She experienced that someone came into her apartment and pressed her down into her bed and she had the idea that she would be cut into pieces. She was also afraid that her former boyfriend would come into her apartment and kill her.

In two years, she had cut her wrists three times and taken pills once. She had been hospitalized ten times. At the hospital, she had been treated with neuroleptics and quickly discharged, but she regularly stopped taking medication at home and had refused outpatient contact. The last time she was hospitalized, she believed she had heard on the radio that she ought to kill herself; she had cut her wrists so deeply that the sinews of her wrist had to be sewn and her forearms had to be put in plaster.

Lotta then agreed to start rhythmic movement training, and I saw her in the ward to show her the exercises. She started to do them every day, and when she was discharged, she told the ward physician that the exercises had helped her to acquire a distance from her thoughts and better contact with herself. She continued to do the exercises two to three times a day for nearly two years and had regular therapy sessions with me. She did not need to be hospitalized again, at least during the next two years before we had to end our sessions because I quit my job.

Lotta's dreams during rhythmic movement training

Lotta dreamed intensely while she was doing the rhythmic exercises. After three months, her dreams became quite violent and she felt very angry. She dreamed that her former boyfriend was dead and that she attended his funeral. In connection with this dream, she started to contemplate suicide and wanted to be admitted to the psychiatric clinic but changed her mind. A few days later, she felt better after having dreamed about a war where people were wounded and killed. In another dream, she was bitten by a snake.

A few months later, she dreamed that a man tried to rape her in a church. She killed him, cut him into pieces, and put him in a plastic bag, which she then carried around. After having done the rhythmic exercises for seven months, she dreamed that her former boyfriend had murdered Swedish Prime Minister Olof Palme and that she would be publicly disgraced when this was known. She therefore decided to commit suicide and was assisted by a doctor who cut her wrists and put on a bandage. After having told me this dream, she added that the last time she tried to commit suicide, she had had the delusion that her former boyfriend had killed Olof Palme; therefore, she had decided to kill herself by cutting her wrists.

For the next few months, she felt good and dreamed several pleasant dreams about being with her mother.

After having practiced the exercises a little more than a year, she got worse and started to feel frightened. She dreamed that people broke into her apartment and strangled her or shot her. These dreams reminded her of what she experienced when she was psychotic, and she now could tell me how she used to crouch in a corner waiting for someone to come in and kill her.

After these dreams, she felt much better and started to dream about being the manager of a big company that was very successful and in which everyone got on very well.

Lotta continued to do the exercises for another year, during which she steadily got better, and after nearly two years, she stated that she had not felt as good for many years.

The dreams as confirmation of recovery from a psychosis

As stated, the prefrontal cortex can be considered a comprehensive operational system, the task of which is to coordinate all our activities, talents, and creative abilities in relationship to the surroundings and to our inner needs. The prefrontal cortex can be compared to the manager or the chief executive officer of the brain and the person. As we have seen, the prefrontal cortex is disconnected in psychosis.

A psychotic person expresses this disconnection symbolically by the idea that the king, the prime minister, or the father has died. This is one of the most common delusions in psychosis. Lotta was only one of many of my patients who soon after the murder of Olof Palme had the delusion that they had killed him. She believed that her boyfriend had killed Olof Palme and therefore tried to commit suicide.

Not until she dreamed about this after having done rhythmic exercises for eight months was she able to tell me about this delusion. The dream permitted a dialogue between the symbolic language of the limbic system and the spoken language of the cortex, and she got perspective on her misconception.

If the father, the prime minister, or the king can be considered the symbol of the prefrontal cortex as the coordinator between thought and feeling, the mother or the queen can be seen as a symbol of the emotions or the limbic system.

After the dream about Olof Palme, Lotta dreamed several pleasant dreams about being with her mother. The pleasant dreams about her mother indicate that Lotta's feelings were less characterized by fear and guilt than they used to be before she started rhythmic movement training. The dream about being the manager of a big company that was successful and in which everyone got on well is a clear indication that her prefrontal cortex was now able to exert inner leadership and that she felt much better emotionally. This she confirmed when she said that she had not felt so good for many years.

Lobotomy as a treatment of chronic schizophrenia

In the 1930s, experiments were made on chimpanzees in which the connections between the prefrontal cortex and the limbic system were severed. The researchers discovered that one chimpanzee that used to have severe temper tantrums became meek as a lamb after the operation.[82]

Some doctors wanted to see if this method could be effective in severe psychotic conditions. This was long before there were effective antipsychotic drugs. The first lobotomy on humans was made in 1935. The method got increasingly popular among psychiatrists when hallucinating and violent psychotic patients were turned into manageable, apathetic, and passive persons without will power of their own. In 1949, the initiator was awarded the Nobel Prize.

During a period at the end of the 1940s, lobotomies were carried out in a steady stream in Swedish mental hospitals. In Sweden, altogether 4,500 schizophrenic patients fell victim to the method. Of these patients, one in six is estimated to have died from the operation. Not until a Swedish doctor in 1951 raised an outcry about the death of six patients in a series of twenty-one operations was there a debate in the media. The method was still used in the 1960s.[83]

Gradually, the devastating and incurable effects of lobotomy were recognized. After the operation, there was a personality change of the patients, and relatives found that they behaved like children. Their interest spans were short, and they were easily distracted. In some of them, indolence was outstanding, whereas others showed a lack of self-control, with explosiveness, talkativeness, and laughter. They lost their ability of empathy and feeling and lost their interest in social affairs, politics, and books as well.

Swedish psychiatrist G. Rylander did a study on patients who had been lobotomized. He found that the more introspective patients were aware that they were unable to feel as before, that something had died within them. An operating room nurse commented on her loss of sympathy with the patients. In his study, Rylander interviewed the relatives of the patients. A mother described how her daughter had lost touch with her deep feelings and become hard. There was a general agreement among the relatives that lobotomy results in an indescribable personality change.[84]

By means of the prefrontal cortex, we "remember our future"

One must ask how an operation that disconnects the prefrontal cortex from the limbic system is effective as a treatment of hallucinating violent psychotic patients. This does not seem to square with the theory that schizophrenia is caused by impaired function of the prefrontal cortex. However, this apparent contradiction can easily be explained.

In psychosis, the patient more or less shuts off impressions from the world around and focuses on inner images and emotions, which often are charged with anxiety. After a lobotomy, the situation is quite reversed. The patient has lost much of her contact with her emotions and inner images and is mainly affected by outer stimuli.

By means of the prefrontal cortex, we can imagine our future based on our present situation and our previous experiences of similar situations. It has been said that the prefrontal cortex enables us to "remember our future." Thanks to this ability, we can act in a long-term rational way to achieve what is important to us and avoid unwanted consequences of our actions. We do not need to be ruled by our impulses like small children. Instead, we can imagine and reason about different alternatives of action and their consequences and examine them against the feelings they arouse.

In stressful situations, we may lose this ability of rational planning and decision making. We may get scared or anxious if we feel threatened. There need not be any noticeable threat. It may be sufficient if we feel criticized, ignored, or see ourselves as worthless failures. If the present situation also reminds us of previous traumatic experiences of a similar kind, our feelings may be kindled to a degree where the prefrontal cortex is more or less disconnected and we lose our judgment and ability of clear thinking. We may become panic-stricken or have a temper tantrum. Normally we will calm down after some time and nothing more will happen, but if we are predisposed to psychotic disease, such reactions may trigger psychosis. When the kindling limbic system sends signals to the dopamine-producing nuclei of the midbrain to reduce their activity, the activity will decrease mostly in the prefrontal cortex and the person may lose his basis in reality.

However, the lobotomized patient can no longer feel frightened, anxious, and threatened as before because the connections between the limbic system and prefrontal cortex have been severed by the operation. And since such a person has lost his ability to "remember the future," he can no longer kindle his limbic system with frightening thoughts. Often his feelings are completely blunted, but in some cases, the patient may still react emotionally in concrete situations. Such emotions, however, will quickly pass.

Release phenomena

In lobotomy, the limbic system will be released from the regulating influence of the prefrontal cortex. This causes release phenomena such as tactlessness, poor judgment, emotional instability, and a tendency for temper tantrums and euphoria, which may be seen after a lobotomy. Symptoms of emotional immaturity, problems with concentration, and difficulty focusing attention are also release phenomena because the prefrontal cortex no longer can govern the activity of the limbic system.

After a lobotomy, the patient functions more like a child, unable to plan and take responsibility. With the inability to imagine the future and the consequences of his actions, the lack of judgment becomes a prominent feature.

Rhythmic movement training in chronic psychosis

In the beginning of the 1990s, I was invited to treat a group of severely psychotic patients in a Swedish mental hospital with rhythmic movement training. Most of these patients had been hospitalized for ten years or more and were heavily medicated. They showed many negative symptoms like indolence, passivity, low interest in those around them, and emotional blunting. Most of these patients agreed to do the exercises daily, assisted by the staff.

Due to fortunate circumstances, we got the opportunity to do a scientific study of the patients who practiced the rhythmic movement training. There was also a control group of similar patients who did not do the exercises. Unfortunately, the study did not start until a couple of months after the treated group had begun the exercises, and the initial changes were therefore not registered.

Despite this circumstance, an evaluation after two years showed that the treated group had improved in several respects compared to the control group. The patients who had done rhythmic movement training showed more interest in those around them and took more part in social activities like occupational therapy and different chores in the ward. They had a greater sense of well-being and were less irritable.[85]

The results confirm the positive effect of rhythmic movement training on the function of the prefrontal cortex and the negative symptoms that are so prominent in chronic schizophrenia.

CHAPTER 14

FOOD INTOLERANCE AND RHYTHMIC MOVEMENT TRAINING

Attention disorder as delayed maturity of the brain

As we have seen, experts consider ADHD to be genetically caused. An alternative explanation would be that for some reason, the brain of the child has not received sufficient stimulation for the neurons to branch off and create new synapses. Everything that obstructs the motor development of the infant and prevents his movements will therefore obstruct the development of the brain. In children who have not had sufficient stimulation of this kind, the maturing of the brain is delayed or impaired. Such impaired development will have many consequences, among others a shortage of transmitter substances, which can appear as attention disorder with or without hyperactivity.

As previously stated, many circumstances such as prematurity, brain injury inflicted during delivery, hereditary factors, vaccinations, electromagnetic fields and microwaves from cell phones, food intolerance, toxicity, or disease may affect the motor development of the child. Such factors may cause the infant to skip important steps of his motor development, in that way hampering his motor development and the maturing of his brain.

A lack of stimulation from those closest to the child, being left alone without tactile or vestibular stimulation, or being forced to spend his time in baby seats instead of moving around on the floor will also prevent the brain from maturing properly.

Causes of hyperactivity in autism

As stated in the chapter about autism, many autistic symptoms are not caused so much by insufficient stimulation as by an inflammatory process of the brain. Extensive immune reactions in certain parts of the brain, especially the cerebellum, have been demonstrated in autism. Another cause of inflammation of the brain in autism is glutamate accumulation because of lack of GABA or inability to transform glutamate into GABA. If this is the case, any kind of stimulation of the brain—be it emotional stress, too many impressions, physical activity, or peptides from the gut because of leaky gut—causes the neurons to continue to fire without restraint because glutamate is not transformed into GABA as it should be. Besides inflammation, this also causes hyperactivity and self-stimulating behavior in children with autism. Typical for autistic children is that they often have difficulty calming down after rhythmic exercises like other children normally do. The fear paralysis and the Moro reflexes are usually active in autism, causing an imbalance between the sympathetic and parasympathetic nerve system, with predominance of the sympathetic system. This causes oversensitivity of sounds, light, visual impressions, touch, vestibular stimulation, smell, and taste, and symptoms like stressful eye contact, Low-stress tolerance, fear of darkness and often severe vestibular sensitivity and a predisposition for motion sickness and poor balance. This is another reason that rhythmic exercises involving the vestibular sense may be challenging for autistic children, who often become nauseated and dizzy, stiffen up, or try to get away.

Similar causes of hyperactivity and emotional reactions in autism and ADHD

Children with ADHD, movement, or speech disorders or learning difficulties may have the same reactions to rhythmic exercises as children on the autistic spectrum. Similar to autistic children, they may become hyperactive after the exercises and even react with self-stimulating behavior. They may also have excessive emotional reactions to the training. These symptoms are signs that they may have problems transforming glutamate into GABA and that they easily become overstimulated.

Therefore, there may be other causes of attention and hyperactivity problems in ADHD than insufficient stimulation of the brain during infancy, as described above. Similar mechanisms that cause autistic symptoms may be present in children with ADHD, who are also exposed to vaccinations, heavy metals, and electromagnetic fields but may be less vulnerable than their autistic peers are. Such causes have been dealt with in previous chapters. In autism, food intolerance, especially gluten and casein sensitivity, causes hyperactivity and self-stimulating behavior. The same is often true in ADHD, learning, and movement disorders. In this chapter, I will mainly deal with food intolerance as a cause of emotional symptoms, neurological disorders, ADHD, and related disorders.

Food intolerance and rhythmic movement training

Intolerance to gluten may cause many symptoms seen not only in autism but also in ADHD, learning, and motor difficulties. In gluten-related disorders, there may be difficulties with attention and concentration due to malfunction of the prefrontal cortex caused by inflammation and malfunction of the cerebellum. Poor articulation and phonological problems also caused by inflammation of the cerebellum may contribute to phonological difficulties in dyslexia. Inflammation of the cerebellum may also cause motor difficulties, poor balance, and ataxia.

Peptides that are produced in the intestines in gluten and casein sensitivity are transferred to the brain and trigger the fear paralysis reflex, causing accumulation of glutamate and inflammation of the brain. Symptoms of inflammation of the brain may be hyperactivity, obsessive-compulsive behaviors, aggressiveness, temper tantrums, depression, anxiety, and psychosis. Such symptoms indicate that the child needs a diet in order to avoid excessive emotional reactions to the training. Such a diet may be crucial in order to integrate the fear paralysis reflex and diminish hyperactivity, concentration difficulties, and hypersensitivity of the senses.

Celiac disease

Gluten sensitivity and celiac disease have become major causes of attention difficulties, hyperactivity, and learning difficulties. These disorders can develop at any age, both during childhood and adulthood. In some cases, they develop when the child starts to eat cereals, but they may develop anytime during childhood and adulthood.

Celiac disease has been recognized since the 1950s. It is a chronic immune-mediated inflammation of the small intestine in genetically susceptible individuals. It is caused by a component of gluten in wheat, gliadin, and similar substances in spelt, rye, and barley, causing an inflammatory reaction in the intestinal membrane and flattening of the intestinal mucosa. It is diagnosed with blood tests for specific antibodies and with intestinal biopsy. Celiac disease has been regarded as a disease mainly with symptoms from the gut, but it is now recognized that there may be no intestinal symptoms in a great part of the cases. The frequency of celiac disease is usually estimated to be around 1 percent. However, it is increasing rapidly, and in Swedish children it has increased five times since the 1990s, from 0.5 percent to 2.5 percent today.

Gluten sensitivity

As stated previously, gluten sensitivity has only been recognized since 2012, when sixteen of the world's leading gluten researchers reached consensus on a new classification of gluten-related disorders. People who show no positive antibody tests for gluten yet have more or less severe symptoms after eating gluten are recognized as gluten sensitive. Gluten sensitivity is believed to be many times more common than celiac disease. Some scientists believe that the incidence of gluten sensitivity may be 30 percent or more.

The symptoms of gluten sensitivity and celiac disease are similar, and therefore the two conditions cannot be distinguished clinically. Symptoms may be abdominal pain; eczema or rash; headache; foggy mind; fatigue; diarrhea; depression; numbness in the legs, arms, and fingers; and joint pain. Even if there are no intestinal symptoms, the intestinal mucosa may be affected, obstructing the absorption of nutrients and vitamins.

There is very little research about how gluten sensitivity can affect our health in the end. Today the only way to diagnose this condition is to observe how health and well-being improve by eliminating gluten. Obviously, gluten sensitivity can cause abdominal symptoms, but it is clear that it can also cause behavioral, emotional, and neurological symptoms. Elimination of gluten from the diet in people who have no markers for celiac disease has been reported to diminish depression and anxiety in many people. In children with ADHD or ADD, a gluten-free diet often improves attention, concentration, hyperactivity, and behavioral symptoms; and in autism, behavioral and neurological symptoms usually improve.

Gluten and neurological damage

There is increasing evidence that the prime target of celiac disease and gluten sensitivity is not the intestines but the brain and the neural system. It has been shown that a gluten-free diet can achieve complete resolution of headaches in patients with gluten sensitivity.

According to Dr. Rodney Ford of the Childrens' Gastroenterology and Allergy Clinic in New Zealand, the fundamental problem with gluten is its "interference with the body's neural networks … Gluten is linked to neurological harm in patients, both with and without evidence of celiac disease." And he states, "Evidence points to the nervous system as the prime target of gluten damage."[86]

The neurological symptoms are believed to be caused by antibodies against gliadin that travel in the bloodstream to the brain, where they cause an immune reaction, especially in the Purkinje cells of the cerebellum—the part of the brain that controls balance and coordination and is essential for speech and executive functions. Damage of these cells may therefore affect language in children both with and without autism and cause late speech development and poor articulation.

These conditions seem to have increased many times during the last ten years. I have seen many such children who have not developed speech or have had more or less severe speech difficulties at the age of five or six. Those who develop speech may have very poor articulation, grammar, and vocabulary. Severe motor impairment, balance, and coordination difficulties are common, especially in children who don´t start to speak. I usually recommend that such children have a gluten- and casein-free diet in order to contain the inflammation of the cerebellum. After a period of diet, I recommend rhythmic exercises, which are always helpful in these situations. It happens frequently that speech improves even before they start the rhythmic exercises.

When should you suspect a gluten-related disorder?

When celiac disease or gluten sensitivity develops in infants and small children, you may see symptoms like diarrhea or constipation. These children may have problems gaining weight, they may cry a lot as babies, and they may have movement disorders. When there is delayed speech, poor balance, and poor coordination—or when a neurological examination has found no cause for a more severe movement disorder—celiac disease or gluten sensitivity should always be suspected.

In older children, abdominal symptoms may often be absent in gluten disorders, and then the diagnosis is easily missed. Even if there are no symptoms from the gut, the absorptions of minerals and vitamins may be compromised. This may cause anemia due to lack of iron or vitamin B-12, which are common symptoms in gluten-related disorders, both in children and adults. B-12 deficiency in adults may also cause damage of the peripheral nerves, most frequently in the legs. The symptoms may be numbness, loss of sensibility, or nerve pain (neuralgia).

Also, children who are easily disturbed and oversensitive to impressions, extremely shy, or afraid of the dark are often gluten sensitive. Depression, irritability, fits of emotion, hyperactivity, oppositional behavior, obsessive-compulsive symptoms, and tics are usually symptoms of gluten-related disorders. Many children with celiac disease and gluten sensitivity are addicted to gluten and prefer to eat bread and pasta.

Celiac disease can appear at any time during childhood, and half of the cases appear after the age of seven. The same is probably true about gluten sensitivity. When a gluten-related disorder develops, you normally notice a deterioration of health and behavior. There may be depression, irritability, phobias, or obsessive-compulsive symptoms. Problems with attention and concentration may surface, and school performance may deteriorate. There may be digestive symptoms like constipation, loose stools, bloating, and pain. The child may start to complain of headaches and migraines. Symptoms like eczema and acne are common.

Casein intolerance

In my experience, intolerance to casein is quite common in ADHD judging from the number of children I have seen with typical symptoms and case history. Symptoms may be:

- allergies and eczema
- asthma
- constipation, bloating, stomachache
- bed-wetting
- temper tantrums
- addiction to casein

Children with casein intolerance often have the following symptoms in their case histories:

- cried a lot as babies
- were allergic to milk

- had infancy colic
- had feeding problems in early infancy
- had sleeping problems in early infancy
- suffered repeated upper respiratory infections, especially ear infections
- had adenoids and swollen tonsils
- were conspicuously slow and inactive as a baby

Casein intolerance is very common in autism, and some autistic children with high-functioning autism may be more sensitive to casein than gluten. Casein, which is milk protein, must not be confused with lactose, which is milk sugar. Lactose-free milk is therefore as harmful as normal milk in casein intolerance.

Casein intolerance may cause similar symptoms as gluten sensitivity, like difficulties with concentration and attention. Even depression, irritability, phobias, and obsessive-compulsive symptoms may appear, although less frequently than in gluten sensitivity. Casein intolerance may also trigger the fear paralysis reflex and cause sensitivity to the senses. The most characteristic symptom of casein intolerance is fits of emotions that may be severe and are more common in casein intolerance than in gluten-related disorders.

Case study

A nine-year-old girl came to me because of reading problems and severe visual symptoms when she read more than a couple of pages. She had no food intolerance, her fear paralysis and Moro reflexes were integrated, and there was no hypersensitivity to the senses. She had severe problems with accommodation and had to struggle with the rhythmic exercises for a long time before her reading improved. Finally, she could read ten pages without any visual symptoms.

However, the next time she visited me, she could not read at all because the letters moved and became blurry and her visual symptoms appeared immediately when she read. I checked her fear paralysis reflex, which now was active, and she had become very sensitive to light and visual stimuli. I asked the girl what she had been doing, and it turned out that her mother had bought an iPhone home to the family. The girl, who had never used a cell phone before, had surfed on 3G for one hour every day. I showed the girl some of the exercises for the fear paralysis reflex, and her mother told me she would lock up the iPhone.

At the next visit, the girl's reading had improved again and she was less light sensitive. However, soon afterward, she started to complain about abdominal symptoms such as stomachache and bloating. I checked food intolerance, and it turned out that she had become gluten sensitive. After eliminating gluten, casein, and soy from her diet, her abdominal symptoms improved, as did her accommodation problems and light sensitivity. Except for the EMFs, there were no other obvious reasons that she should have developed gluten sensitivity, such as infections or vaccinations. She still would get a stomachache when she was exposed to high levels of EMFs.

CHAPTER 15

WHAT IS DYSLEXIA?

The expert definition of dyslexia and reading and writing difficulties

According to a consensus statement by twenty-two Swedish dyslexia researchers, one should distinguish between dyslexia and general reading and writing difficulties. Dyslexia is considered to have a basis of language biology. Children with dyslexia "do not succeed in managing the sounds of the language in reading and writing." According to the statement, such difficulties often have a genetic basis.[87]

Reading and writing difficulties, on the other hand, are caused by deficient requirements for language development in the child's family and school, apart from the problems with language biology. All other theories about dyslexia are promptly brushed aside by the dyslexia researchers, who also state:

"Widespread ideas about dyslexia as a consequence of visual disorders, defective motor development, lack of certain nutrients, or deficient interplay of the ability of the brain to process sound signals have no support among the consensus researchers".

The best way to help children with reading and writing difficulties

An ever-increasing number of Swedish children finish the nine-year obligatory school without having learned to read and write properly and therefore do not obtain a pass in Swedish or English. One important cause of that, according to the consensus researchers, is the incompetence of the teachers who do not know how to teach children to read. However, the measures that the researchers propose (i.e., intensified reading instruction) has so far not been effective, and the problem is becoming more and more acute. In fact, 11 percent of the students of the nine-year compulsory school failed to pass to high school in 2008.

When the school fails to teach children to read, many parents have chosen to seek help outside school and discovered that such help may be both quick and effective. By way of movement training, reflex integration, massage, or sound stimulation, many children have improved their reading ability, enabling them to keep up with the schoolwork.

The statement of the twenty-two dyslexia researchers must be seen in this context. In the worst academic tradition, they are entirely focused on proving the correctness of their own theories and dismissing alternative methods that in practice have turned out to be effective. Obviously, they want to prevent unorthodox methods, which can make it easier for children to read and write, from being used in Swedish schools. In my experience, they have been successful in preventing school leaders from trying such methods that could easily help most children with reading and writing difficulties.

In the following chapter, I comprehensively deal with the causes of dyslexia that the twenty-two researchers so hastily dismiss. My purpose is to explain why motor training and reflex integration works so much better in reading difficulties than intensified reading instruction and speech training.

The sensory process

Reading and writing is a complicated process that includes many subordinate processes. In order for us to be able to read and write, there must be a sensory process that receives impressions (letters, words, or signs) and transfers them to specific areas of the brain in the neocortex. By means of the perception process, we become aware of, or perceive, sensory input. Many different areas of the neocortex must work together in a neural network in order for the reading process to function. For the sensory process to function properly, our sense organs and brains have to work adequately. A proper functioning of the eyes is not enough for good vision. The eyes must be able to work together (binocular vision) and move properly, and they must be able to change the focus of vision from near to far distance (accommodation). All these abilities are controlled from different centers on various levels of the brain, which have to collaborate. As will be shown, our motor abilities are fundamental for visual skills such as accommodation, binocular, vision and eye movements.

The perception process

The perception process depends on the functioning of the sensory process. Impressions from the senses are processed in different areas of the neocortex in order for us to become aware of them.

Visual impressions are processed in the occipital lobe, auditory impressions in the temporal lobes, and sensory impressions in the front part of the parietal lobe. If the sensory process does not work properly, the corresponding parts of the neocortex will not be sufficiently stimulated, which will obstruct the proper functioning of these areas and afflict the perception process.

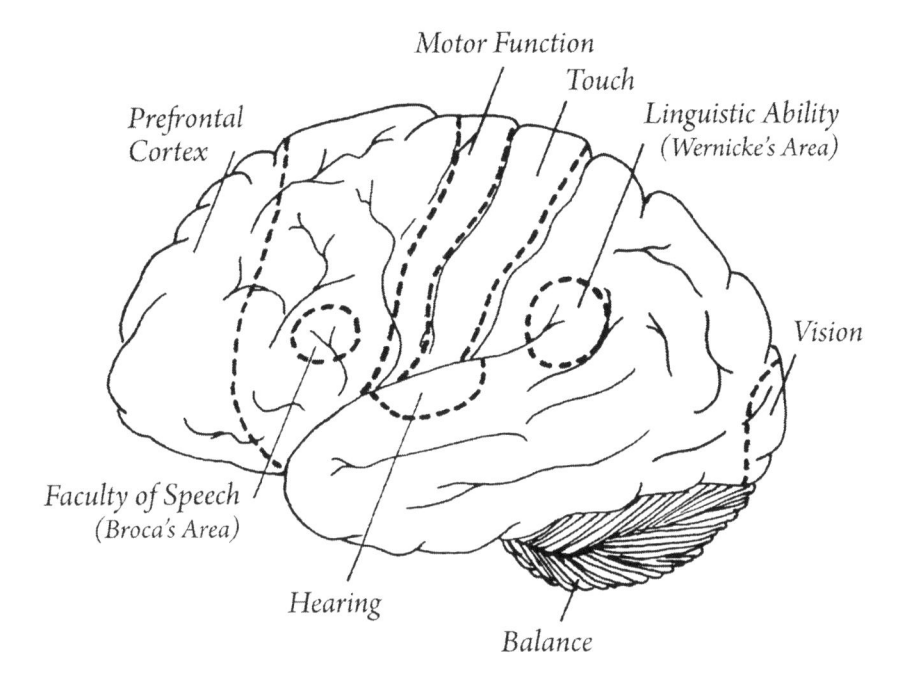

Fig. 21 Areas of the cortex important for reading

The perception process also depends on our alertness and the arousal of our neocortices. When we listen to a boring lesson, the arousal of our neocortices decreases and we may start daydreaming. We will no longer perceive the stimuli from the world around us that are processed in the neocortex. Instead, we will be aware of memories and imaginations mainly processed in the limbic system. We may hear the lecturer speak, but we don't know what he is saying.

In a similar way, we may be reading a book aloud and suddenly discover that we have been thinking of something else and haven't a clue about what we've been reading. In this case, the reading process has been entirely automatic.

It may also happen that while listening to a lecturer speaking in a foreign language or on a complicated abstract subject, we have great difficulties understanding what the lecture is about and have to work hard to decode the meaning of it. Although we try to stay focused and concentrated, we may soon tire and give up understanding what the lecture is about and start daydreaming.

Being able to read means that reading has become an automatic process

A child who has not learned to read properly is in a similar situation as to what has been described above. The very reading process, decoding letters and words and perhaps sounding out the words laboriously, may require so much of the child's attention that he or she really hasn't got a clue what the text is about. In the practiced reader, this part of the reading process is completely automatic and the person's perception can instead be focused on comprehending the text.

There is a great difference between the adult who is thinking of something else when reading text and a child who hasn't yet learned to read properly and is struggling to understand the meaning of the text. The adult reads automatically but does not grasp the meaning because he or she is thinking of something else. The child who has not learned to read properly and is struggling to understand the text has not yet learned to coordinate the neural network of reading and the automatic reading process.

Most children gradually learn how to read by more or less assiduous practice, as most people learn how to drive a car by practice. Some children, however, have great difficulties in coordinating the subordinate processes of reading and therefore do not attain the level of automatic reading.

Different causes of reading and writing challenges

Common to all children with reading challenges is their inability to make reading an automatic process in spite of intensive practice. Researchers have previously proposed different theories about the causes of such an inability. But their theories have become obsolete among dyslexia experts, and dyslexia is nowadays considered to be due to one solitary factor: a weakness of the sound aspect of language or a phonological disability. According to Swedish experts who share this outlook, nothing can be done about dyslexia except to intensify the practice of reading.

Another way of tackling the problem of reading and writing disability is to proceed from the individual child and find out which specific challenges may cause the difficulties. Modern research has shown that various centers both in the neocortex and on other levels of the brain must cooperate in a neural network for the child to be able to read and write. There is no simple answer to the question as to why this network does not work properly in a child with a reading disability. The causes are different in different children. With various methods, it is possible to get an understanding of the underlying causes that obstruct reading and writing in the individual child and learn how to stimulate these abilities.

Without sophisticated methods to evaluate the activity in various parts of the brain, such as PET (measuring the blood flow of the brain) or an MRI, one must draw conclusions in other ways about the causes of a reading disability. Primarily, it is important to conduct a thorough interview to get an understanding of the nature of the reading disability, how it manifests, and what subjective symptoms there are. By examining motor ability, reflexes, and vision, which can be done by simple means, it is possible to reach an understanding of the nature of the reading disability and learn how to proceed to stimulate the neural processes and improve reading ability.

Various theories about the causes of dyslexia

Dyslexia researchers have found various subordinate groups of dyslectic children. By analyzing the ability of reading and spelling in dyslectic children, Boder[88] classified dyslectic children into three groups. According to Boder, the most common problem is difficulties with phonetic analysis. Children in this group read words as whole images. Their spelling skills are poor, and since their reading is based on recognition, they have great trouble reading unknown words and often make wild guesses when reading. According to Boder, these children suffer from dysphonetic dyslexia and constitute nearly two-thirds of dyslectic children.

Children who manage phonetic analysis but have poor visual memory belong to a smaller group, around 10 percent. They sound out the words laboriously, as if they are always seeing them for the first time. They spell poorly, especially if the spelling is irregular. This group is called dyseidetic dyslexia.

A third combined group constitutes the hard core of dyslectic children. They have challenges with both phonetic analysis and visual memory. They frequently have difficulties in learning the name of the letters and memorizing their appearance. They often make transpositions and reversions of letters. Unlike the other two groups, whose prognosis is better, their reading and writing challenges often remain into adulthood. They constitute about 20 percent of all dyslectics.

A similar classification has been made by Gjessing.[89] He distinguishes one group with auditory challenges corresponding to the dysphonetic group of Boder. They have challenges in distinguishing similar sounds like *b* and *p*, and they may guess frequently when they read. He labels this group auditory dyslexia.

Gjessing also distinguishes a group of visual dyslexia corresponding to the dyseidetic group of Boder. They have challenges in perceiving the words as images and completely rely on sounding out when they read. They spell the words according to how they sound, and they omit silent letters.

Sequential and simultaneous processes

The classification of dyslexia according to sense modalities has been criticized. Critics have emphasized that dyslexia rather is a matter of two different process modalities, the sequential and the simultaneous, rather than different sense modalities.

In the simultaneous modality, words are perceived as images or visual totalities. The sequential process is characterized by phonetic analysis.

Spoken language is sequentially organized, while written language is simultaneously organized. Both these processes must operate in order for reading to be fluent and automatic. We use the simultaneous process to perceive the meaning of the word image and the sequential process to understand the meaning of the text.

In alphabetic languages, there are two ways of perceiving the meaning of a word. If we are familiar with a specific written word, we can directly understand its meaning by recognizing the word image. If not, we can analyze the syllables phonetically and sound out the word in order to hear if we recognize it.

In a language like Chinese, where each word corresponds to a pictorial sign, it is impossible to analyze the words sequentially and words must be memorized as images.

Aaron[90] has demonstrated that children who are diagnosed with dysphonetic dyslexia have challenges with the sequential process and children with dyseidetic have challenges with the simultaneous one.

The specialization of the hemispheres of the brain

The two hemispheres are specialized in different tasks.[91] It has long been known that the left hemisphere is specialized in linguistic tasks, such as analyzing the sounds of the language (phonetic decoding), to speak and understand spoken language and to use grammar and understand the meaning of language. The right hemisphere is more specialized in comprehending context, understanding metaphors, and perceiving intonation and emotional undertones.

In the visual area, the right hemisphere specializes in seeing the whole picture (e.g., faces) and in seeing details in their spatial context. Individuals with right hemisphere injury who make a drawing of an object may include details, but these are not coherent in a meaningful way, and these persons also have difficulties depicting three-dimensional aspects.

Two areas are especially important for language, namely Broca's area in the left frontal lobe and Wernicke's area in the left temporal lobe. Injuries to Broca's area may cause an inability to speak, which is called expressive aphasia. Such injuries also impair the ability to utilize and understand the grammatical structures of language.

Injuries of Wernicke's area cause challenges in expressing oneself in a meaningful way and in understanding the meaning of spoken language.

Studies have shown that 96 percent of all right-handed people have their speech centers in the left hemisphere. In the terminology of the researchers, that means that the left hemisphere is dominant. In the majority of left-handed people, the left hemisphere is also dominant: the speech center is situated in the left hemisphere in 70 percent of the left-handed.

In 15 percent of the left-handed people, the speech center is found in the right hemisphere, and in another 15 percent, there is no preference (that is, both hemispheres are considered dominant).[92]

In the traditional notion about the hemispheres, sequential processes are primarily processed by the left hemisphere and simultaneous processes primarily by the right hemisphere.

However, research during the past few years has thrown a new light on the specialization of the hemispheres.[93] Using PET scans (measuring the blood flow of the brain), it has been shown that unfamiliar phenomena, which can be words, objects, or symbols, activate both the right and left hemispheres. The more the experiment is repeated, the less activation there is of the right hemisphere, while the activation of the left one remains unchanged. The same thing happens when we are exposed to faces, commonly thought to be processed in the right hemisphere. Experiments have shown that every time we see an unfamiliar face, both the right and left hemisphere are activated, but for every additional exposure, the activation of the visual cortex of the right hemisphere will decrease, while that of the visual cortex of the left hemisphere will increase.

From this research, one can draw the conclusion that simultaneous processing takes place in both hemispheres. When we are exposed to unfamiliar phenomena, the right hemisphere is more important than when we are exposed to familiar phenomena.

The specialization of the hemispheres serves an important purpose by enabling different parts of the brain to fulfill various tasks as has been described above, which increases the efficiency of the brain. If there is poor specialization of the brain, the neural process of the brain will be more global, and complicated neural processes, like reading and writing, will require more effort. Consequently, poor specialization of the brain may be one cause of reading and writing difficulties.

CHAPTER 16

VISUAL CHALLENGES AND DYSLEXIA

Visual challenges and reading challenges

According to modern dyslexia theories, visual challenges do not cause dyslexia; instead, it is more defined as a condition caused by phonological challenges.

However, researchers who have investigated the relationship between reading disability and visual challenges have estimated that visual challenges are an exclusive or contributing cause of reading disability in about 40 to 50 percent of cases. In spite of that, there are ophthalmologists and opticians who do not believe in the relation between visual challenges and reading disability. Due to this circumstance, many children whose reading challenges would easily be improved or remedied by prescription of suitable glasses do not get any help.

Moreover, it is probably unreasonable to make a definite distinction between children with phonological challenges and those with visual challenges, especially with the hard core of dyslectic children who have both phonological and visual challenges. In addition, visual challenges that cause the left eye of a right-handed child to become dominant may even cause phonological challenges when the child learns to read.

Development of vision and visual skills

Vision—the ability to perceive, identify, interpret, and understand what we see—is not an inborn skill but is developed and learned from infancy up to the age of twelve years. The development of vision and motor ability is interrelated. By means of the inborn program of baby movements such as grasping, putting things into his mouth, lifting his head whilst in a prone position, crawling on his stomach, and crawling on hands and knees, the baby learns to develop his visual skills. Babies who do not make these baby movements satisfactorily because of motor challenges may develop visual challenges. Infants who do not learn to grasp thing and put them in their mouths will have no opportunity to practice eye-hand coordination and will not exercise binocular vision or fusion. Children who do not learn to crawl on hands and knees do not practice the ability to move the eyes from short to long distance and may develop challenges with accommodation.

Binocular vision and accommodation are skills that should be established during the first year of life, while adequate eye movements and eye dominance requires that the two hemispheres cooperate satisfactorily. The cooperation between the hemispheres is initiated by the integration of the asymmetrical tonic neck reflex (ATNR) and is developed when the baby crawls on her stomach and hands and knees. An adequate cooperation between the hemispheres does not come about until there is a proper myelination of the corpus callosum, which does not occur until much later. That is why the ability to continuously follow a moving object with the eyes without their stopping in the midline or making irregular jumps does not develop until the age of seven or eight years and total eye dominance is not established until ten or eleven years of age.

Visual skills such as binocular vision, accommodation, and eye movements are important for learning how to read and for the further development of the reading ability.

Binocular vision

The ability to integrate the visual images from our two eyes into one image is an important visual skill. One condition for this is our ability to direct our eyes at the object we are looking at, which is called vergence. Since the eyes are a bit apart, the images of this object that end up on the central visual field of the eye or the fovea are somewhat different. The eyes see the objects from different angles. These two images should then be fused into one three-dimensional image in the visual cortex, which is called fusion.

If the individual is not able to direct the eyes in such a way that the image of the observed object ends up in the central visual field, binocular vision will be impossible and be replaced by double vision or suppression of one of the eyes. If one eye is constantly suppressed, the visual acuity of that eye will decline. This condition is called strabismus. Occasional suppression that only occurs when in visual stress (e.g., reading) is called suspension.

Phorias

In some individuals, the direction of one or both of the eyes will change when the eyes relax. Unlike persons with strabismus, they can still direct both eyes at the same point in space when they look at it; they usually have no binocular challenges like double vision and suspension.

At rest, the eyes tend to be directed inward, esophoria, or outward, exophoria. More unusual is hyperphoria, when one eye is directed more upward or downward than the other one.

Unlike in strabismus, binocular vision normally functions in phorias; however, the more marked the phoria is, the more strain there will be on the eyes. In persons with marked phorias, binocular vision may sometimes require such a big effort that they have headaches in their foreheads or around the eyes when they become tired or stressed. They may also experience challenges such as pain or tiredness in the eyes.

Children who have marked phorias and challenges with binocular vision and who have not yet made the reading process automatic often get headaches either in their foreheads or in the backs of their heads, or irritation of the eyes when they read. They lose concentration, they read slowly and with great effort, and they have challenges understanding and remembering what they read. When they get tired, demands on binocular vision are greater than they can manage and there will be double vision or suspension—or even suppression of one eye. Phorias may be found in vision both at near and far distance and sometimes only in vision at near or far distances.

Accommodation and myopia

Our vision is not constructed for work at a close distance but mainly for good visual acuity at a long distance in order for us to find food and discover threats around us. Our modern culture, however, makes quite different demands on our vision. From an early age, our eyes must get used to working at a close distance, and in some cultures, children are taught to read at the age of four or five. When we look at something at a distance, the image normally ends up on the retina. In order to move the focus of our eyes from an object at a distance to one close to us, we must accommodate, which occurs by the lens changing its form and becoming more spherical. The lens is elastic and suspended by thin threads on the surrounding ciliary muscle. When the ciliary muscle is relaxed, our vision should normally be adjusted for vision at a distance. When we accommodate, the ciliary muscle will contract and the lens will change its form and become more spherical. In children, the lens is still very formable, but after the age of forty, it loses its plasticity and does not change its form in the same way when the ciliary muscle is contracted. Then our powers of accommodation diminish and we have to get reading glasses.

In accommodation, when the ciliary muscle contracts and the lens becomes more spherical, the pressure within the eyeball increases. If there is recurrent accommodation due to frequent work of the eyes at a close distance (like reading), the ball of the eye will grow longer due to the increased intraocular pressure. In a way, such an adaptation is appropriate since it demands less accommodation and strain on the eyes and especially on binocular vision during work at near distance. But we have to pay a price for this adaptation because when we look at an object at a distance, the image will no longer end up on the retina but in front of it; therefore, it will be blurred. In this way we develop myopia and have to correct our impaired vision by using concave minus glasses when we want to see clearly at a distance

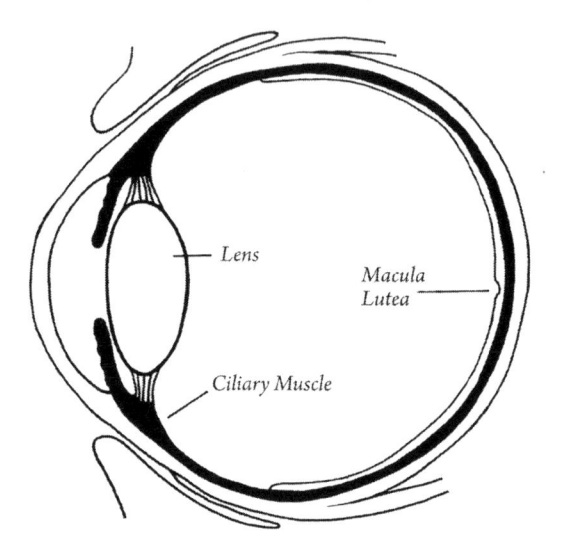

Fig. 22 A section of the eye with the lens, ciliar muscle, and macula

A contributing cause of myopia is deficient accommodation and an inability to change the focus of the eyes from near to far distance, and vice versa, instantly. For children with this problem, it might take ten to twenty seconds or more before they can see clearly when looking from the board to the book. The more such a child accommodates, when reading or working at a short distance, the less opportunity the ciliary muscles will have to relax. This problem might even cause the ciliary muscle to go into spasm, creating a condition of pseudo-myopia. In this case, the myopia is not caused by the prolongation of the eyeballs but by the constant accommodation of the lenses.

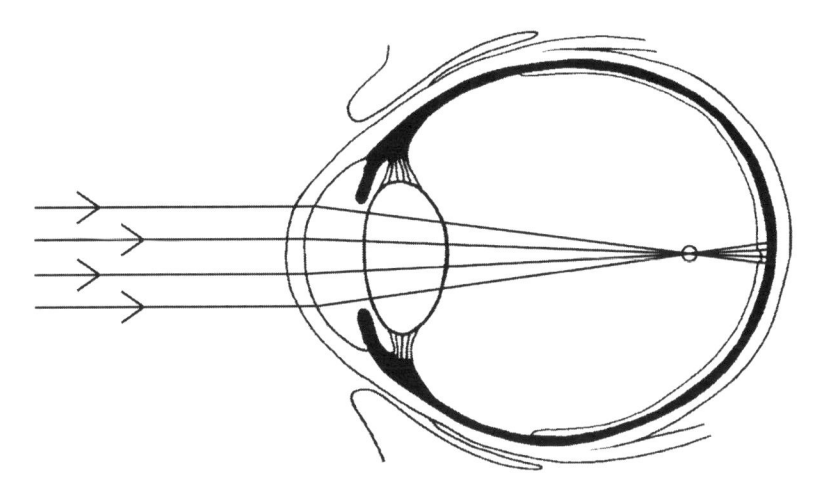

Fig. 23 Myopia

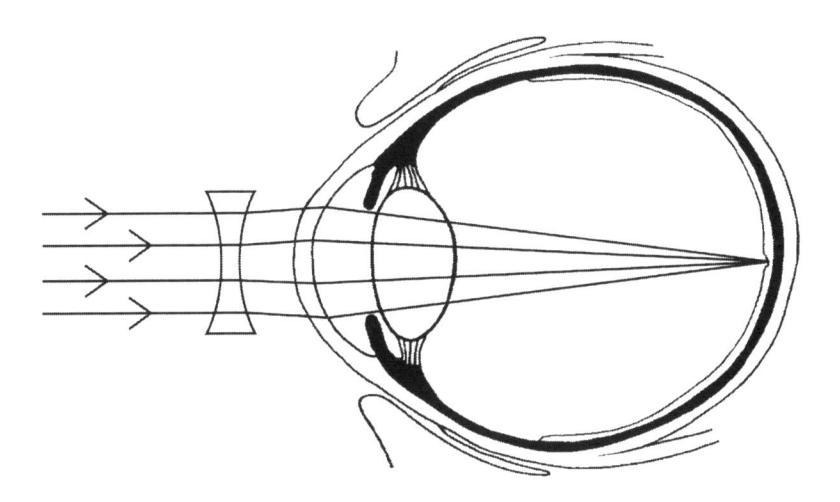

Fig. 24 Correction of myopia with concave lenses

The adequate treatment of this condition is to give the child reading glasses to make the ciliary muscle relax. In this way, the development of myopia may be prevented. Myopia does not occur in peoples who do not have a written language.

Hypermetropia, accommodation, and convergence

Small children normally are slightly farsighted (hypermetropic). This means that the eyeballs by nature are a little too short so they have to accommodate slightly in distance vision in order for the image to end up unblurred on the retina. Usually this is of no practical importance since small children have a good ability to accommodate. However, some small children may be very farsighted, causing the image to end up behind the retina and become blurred even if they accommodate as much as they can. Then they need concave plus lenses both at long and short distance in order to see clearly.

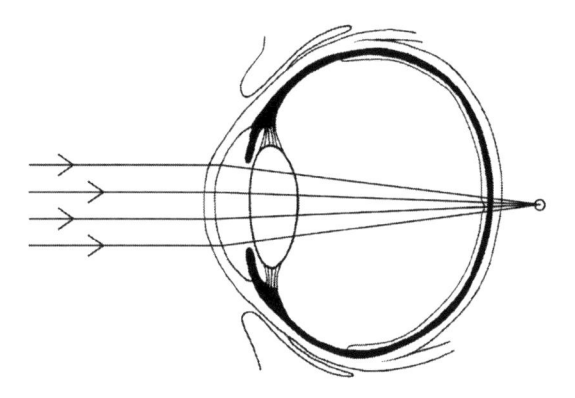

Fig. 25 Hyperopia

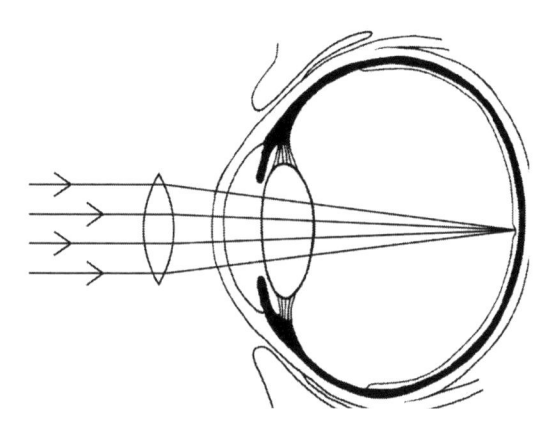

Fig. 26 Correction of hyperopia with convex lenses

The cause of such striking farsightedness is not known for certain. One explanation could be that these children, because of some motor handicap, have not used their hands sufficiently to practice eye-hand coordination. Because of insufficient practice of accommodation, the vision of these children will mainly be adjusted for seeing at a distance. Perhaps the practice of accommodation in normally developed children is necessary to counteract the increase of inborn farsightedness. If, in addition, the accommodation is slow, which will certainly be the case in most motor handicapped children, the effort of accommodation will increase, which will further diminish the inclination to accommodate.

When we accommodate, there is an automatic change of the direction of the eyes by means of the outer eye muscles, causing the eyes to be directed inward or converge, which is called accommodative convergence. The more we accommodate, the greater will be the convergence.

Children with a slight farsightedness manage to see clearly at a distance if they accommodate to compensate for the short eyeballs. When these children look at a short distance, they therefore must accommodate even more, thus straining not only their accommodative faculties but also their convergences. If they have deficient accommodations, the effort to maintain binocular vision will add to challenges with accommodation and cause headaches over the forehead and in the back of the head as well as exhaustion and smarting pain of the eyes. When they read, they get tired, and they usually manage to read only for a short time. When they do not manage to maintain binocular vision, they will either develop double vision or suspend one eye.

When these children are examined with an instrument, there will be a more or less marked esophoria at reading distance because of the accommodative convergence. This esophoria is usually corrected if the child is given reading glasses, which decreases the accommodation and therefore also the convergence.

Saccadic eye movements

When we read, the eyes move by jerks along the line, which is referred to as saccadic eye movements. In the cortex, they are initiated by the frontal eye field. This area of the frontal lobes has important neural connections with the cerebellum. With every jerk, the eyes move from one fixation point to the next one. An unpracticed reader makes more and longer fixations than a practiced one. In addition, unpracticed readers more often move backward along the line, so called regressions. The same is true for dyslectics.

We cannot consciously control the speed and accuracy of these movements. If they are slow, reading and schoolwork will be slow, and if they are inaccurate, the child will make careless errors.

One or more of the following signs could suggest a problem with saccadic eye movements:

a. The child moves his head rather than his eyes from side to side when reading.
b. The child frequently loses his place and skips lines when reading.
c. Past the age of seven, the child needs a finger to keep his place on the line.
d. The child may alter or miss beginnings or ending of words or skip small words.
e. The child may be labeled as having attention problems.

Pursuit eye movements

Pursuit eye movements, also called tracking, are the ability to follow a moving object with the eyes. The pursuit eye movements should be gentle and continuous without accessory movements of the head. They can easily be tested by asking the client to follow your pen with his eyes when you draw a big *H* in the air. You should especially notice if the eye movements are jerky or twitch or stop in the midline. The ability to track in a smooth and continuous way usually develops at the age of seven or eight years and is dependent on myelination of the corpus callosum and cooperation of the hemispheres. Both the frontal lobes of the neocortex and the cerebellum are important for the pursuit eye movements. There seems to be a connection between poor pursuit eye movements and reading disability in the sense that poor pursuit eye movements reflect poor cooperation between the hemispheres, which obstructs reading.

Eye dominance

When we are faced with a choice, we choose to use one particular eye, the dominant one. It is easy to decide which eye is dominant by means of a piece of paper with a hole in the center (about one inch). It is important that the client hold the paper with both hands at a distance of forty to fifty centimeters from the eyes. The client is then asked to look at a pen or some other small object first at a long distance, about five meters, and then close to the face. The client is asked to close one eye at a time without moving the paper, and the eye that sees the object is the dominant one.

When children learn to read, they normally have not developed a dominant eye. This does not happen until the age of ten or eleven. The part of our visual field to the right of the fixation point ends up in the left hemisphere. Since the right eye is dominant for the right visual field, the left hemisphere will mainly process information from the right eye and the right hemisphere will process information from the left one.[94] Cooperation between the hemispheres is necessary when we read. With the right hemisphere, we see the words as whole images, and with the left, we analyze the words and letters, especially phonetically. If we only use one eye when reading and suppress the other, the corpus callosum will be single-handedly responsible for this necessary cooperation. In binocular vision, on the other hand, we can use the eyes to choose which hemisphere to activate.

At the age of starting school, the connections of the corpus callosum between the right and the left hemisphere do not function very well. The children who do best when learning to read are those who can change eye dominance according to need. By reading with the left eye (i.e., the right hemisphere), they see the word as a whole image and learn to recognize it. When they meet totally unfamiliar and difficult words, they change to the right eye and left hemisphere and analyze the words sequentially and phonetically.

Children who have developed a strong dominance of one eye before they learn to read because of problems with binocular vision may have problems learning to read well. If the right eye is heavily dominant, they have no challenges learning how to read by sounding, but they will have challenges achieving swiftness of reading since they do not learn to recognize the words as pictures since the right hemisphere is not stimulated in reading.

If the left eye is heavily dominant, mainly the right hemisphere will be stimulated. These children will have challenges in being able to analyze the words and learn to read by sounding, and therefore they will develop phonological challenges. They easily learn to recognize the words but will have challenges in decoding unfamiliar words.

When the connections between the hemispheres develop at the age of ten to eleven years, the cooperation between the hemispheres will come about more by communication through the corpus callosum and not by changing eye dominance. It is also at this age that eye dominance is normally established. At this age, many children learn to compensate by means of the corpus callosum for binocular challenges, causing eye dominance to develop prematurely. If up until then, they have had challenges reading, their reading ability now will develop quickly. Moreover, the dominant eye controls how the eyes follow the lines. The right eye naturally moves from left to right, while the left eye moves from right to left. Learners who are dominant with the left eye will start to look at the right side of the page first and will have more challenges reading languages that are written from left to right, such as English. These learners have a tendency to make reversals of letters and mix up *b* and *d*. Sometimes they prefer to have the text upside down when they learn to read.

Challenges with accommodation that affect reading

Nearly all children with dyslexia have challenges with accommodation. Besides binocular problems at a close distance, such challenges may affect reading ability in other ways. There may be problems with flexibility of accommodation. In such cases, it may take several seconds, sometimes half a minute or more, to see clearly after changing focus from a short to a far distance. There may also be challenges with stability of accommodation. In such cases, the child is unable to keep focusing on the text. If the focus is not stable, the text will become blurry. These problems usually become evident after the child has been reading for a while and reading becomes more challenging the longer the child reads. The child needs to strain his eyes to see clearly, which may cause headache and irritated eyes. To compensate for poor stability of accommodation, the child may move the text back and forth to see more clearly. Signs of accommodation problems:

- The child tires soon when reading, and reading comprehension diminishes the longer reading is continued.
- The child avoids reading or reads as little as possible.
- The child gets headaches or irritated eyes when reading. He may start to rub his eyes after reading for a while.
- The child blinks excessively when reading or looking at street signs in order to see more clearly.
- The child complains that the text becomes blurry.
- The child holds his book too close to his eyes or moves the book or his head closer and farther as if to clear up things.
- The child makes careless errors when reading or copying from the board. Little words such as *of, as*, and *is* or small beginnings and endings of words are misread, while long words like *hippopotamus* are recognized.[95]

Challenges with accommodation due to stress

Accurate accommodation is usually dependent on the integration of the STNR. However, the fear paralysis and the Moro reflexes are also very important. The ciliary muscle, which controls accommodation, is innervated by the parasympathetic nerve system, which needs a tranquil and relaxed frame of mind to function well. In stress (e.g., when the Moro or fear paralysis reflexes are triggered), the sympathetic nerve system will take command and obstruct contraction of the ciliary muscle. When the ciliary muscle cannot contract, accommodation will be obstructed. The lens will become flat, and the eyes will focus at a far distance. When the fear paralysis reflex is excessively triggered in extreme stress and the individual becomes paralyzed with stress, the ability to see clearly even at a far distance may cease and everything may become blurry.

In some individuals with an active Moro reflex and severe sensitivity to light, the contrast between a bright paper and black text may obstruct accommodation because of sympathetic stimulation and cause the text to become blurry and the letters to move and sometimes even disappear completely.

Visual perception

Visual perception includes the ability to compare things and see how they are different. In order to be able to learn letters, a child must be able to perceive the difference between a circle, a square, and a cross. In order to tell the difference between *b* and *p*, the child must have some understanding of the concepts of *top* and *bottom*. And in order to tell the difference between *b* and *d* the child must understand what is right and left.[96]

Visual perception gradually matures during the first four or five years of life. This maturing can be promoted by doing rhythmic movements, which will stimulate the cortex and consequently perception. Visual perceptions can also be practiced by letting the child draw figures like circles, squares, and so forth. Perceiving the difference between *b* and *d* may still be a challenge for some children when they start school, and many of these may develop difficulties with reading and writing. Challenges with left and right may also cause frequent reversals of letters in reading or writing (for example, *saw* becomes *was*). Such difficulties are usually linked to a retained ATNR reflex, a reflex that is of fundamental importance in dyslexia.

Visual challenges need not affect reading ability

Certain children with marked visual challenges like strabismus or challenges with accommodation or binocular vision may still be able to read quite well in spite of their challenges. This fact has caused some dyslexia researchers, oculists, and opticians to deny the connection between visual challenges and dyslexia. It is certainly not easy to tell just from a visual examination if a child suffers from a reading disability. However, studies comparing groups of good and poor readers have demonstrated which visual challenges predispose reading disability. Challenges with accommodation, farsightedness, suppression of one eye working at a close distance, poor binocular vision with alternate suspension, and poor tracking are significantly more common in poor readers than in good ones. But many children do not experience any trouble and are able to read for quite a while without getting tired, in spite of one or more of the above-mentioned challenges. By practice, these children learn to make reading automatic and to compensate for their visual challenges.

The consensus statement by twenty-two Swedish dyslexia researchers categorically denies that visual problems could have anything to do with dyslexia. They base their opinions on extensive studies showing that binocular problems are not more common among people with dyslexia than among normal readers. This argument not only shows how little these scientists know about the challenges of dyslectic children, but it also demonstrates how they have abandoned all common sense in order to pursue their phonological theories. Otherwise, they would understand that children will find it difficult to read and easily tire when they experience irritation or smarting of the eyes or headache—or the text becoming blurry or starting to jump. If they could overcome their denial and instead ask children with dyslexia about such visual problems, they would find that at least 80 percent of them have one or several of these challenges.

These children are not difficult to discover. They are often able to read only for a short while, and they will cover one eye with one hand or hold the text very close to the eyes. When asked, they may tell about double vision or the text becoming blurred after a short period of reading. They may also admit that they get headaches in the forehead or in the back of the head after reading for a while or that the eyes become irritated or start to water whilst reading. Children with these challenges read slowly and unwillingly and often have difficulties understanding what they read. However, if nobody asks them, they will not complain about their symptoms. They are convinced that these symptoms are normal for every child. To recommend increased reading practice to these children, as the experts do, is nothing less than an insult. Reading practice will rarely improve their reading. Even more, it will convince them that they are hopeless cases.

The development of binocular vision during infancy

As has been mentioned the developments of vision and motor abilities are interrelated with each other. By means of the inborn program of baby movements such as grasping objects, putting them into one's mouth, lifting one's head in a prone position, crawling on one's stomach, getting up on hands and knees, rocking and crawling on hands and knees, and so forth, the infant learns to develop its visual skills. Children who do not make these movements adequately because of motor challenges usually show many challenges with vision.

By means of the outer eye muscles, we can see in different directions. In order for us to have functioning binocular vision, we must be able to direct our eyes so that the image of the object we are watching ends up on the central visual field or the fovea of *both* eyes. This ability implies cooperation between the two eyes. Since the eye muscles that direct the right eye are controlled from the left hemisphere, and vice versa, the two hemispheres must cooperate in order for binocular vision to be possible.

In newborn babies, the hemispheres do not cooperate. Due to the asymmetrical tonic neck reflex (ATNR), the movement pattern of the newborn infant is mainly homolateral, which means that the body halves do not cooperate but move separately. When the infant turns its head to the right, the right arm and legs are stretched, while the left ones are bent. Likewise, the left arm and leg will stretch while the right ones will bend when the baby's head is turned to the left. During the first weeks of life, the infant spends 80 percent of its time in such an ATNR position. It is an advantage if the head is turned to the left, for in this way, right ear dominance is naturally established.[97]

When the infant lies on his back or stomach with his head in either direction and learns how to fixate his eyes on his hand or an object, how to grasp the object and bring it to its mouth, eye-hand coordination and binocular vision are developed. Later the baby will learn how to move the object from one hand to the other, which indicates that the hemispheres are learning to cooperate.

Probably the most important primitive reflex for the development of hand-eye coordination and binocular vision is the asymmetrical tonic neck reflex (ATNR). Other reflexes are also very important. The grasp reflex enables the baby to grasp things and to release them and move them to the other hand when the grasp reflex starts to integrate. When the baby grasps something with her hand, the Babkin reflex is activated. Sucking movements of the mouth will stimulate the baby to put the object in her mouth, and the hands pulling reflex will help her to bring it to her mouth.

The ATNR, the grasp reflex, the Babkin reflex, and the hands pulling reflex belong to the reflexes that develop the midline dimension (i.e., the cooperation between the hemispheres, the body halves, and the eyes). If these reflexes have not developed normally and have not been properly integrated, there is an increased risk of developing strabismus or other challenges with binocular vision.

The integration of ATNR and other midline reflexes is important for pursuit eye movements. When these reflexes are integrated, the cooperation between the hemispheres and the body halves improves and more nerve signals pass the corpus callosum, which increases the myelination of its nerve fibers. All kinds of cross movements the child makes, like crawling on the stomach or hands and knees, also stimulate the myelination of the corpus callosum. If the child has not integrated the ATNR and/or has not crawled, it may have challenges crossing the midline. The pursuit eye movements may become jerky and may stop in the midline or the child may blink in the midline.

The importance of integrating the ATNR in dyslexia

In light of the account above, it is not difficult to understand why the integration of ATNR is so central in reading and writing difficulties. The effect of integrating the ATNR can be summarized in the following way:

- The integration on the ATNR stimulates the myelination of the corpus callosum, improves the communication between the hemispheres, and increases the speed of transmission in the neural network of reading.
- It improves binocular vision and tracking eye movements.
- It improves the motor ability of the hand and the arm and improves writing ability.

The importance of the ATNR in dyslexia has also been confirmed by research. An article in *The Lancet*,[98] published in 2000, reports about a study on the effect of a retained ATNR in dyslexia.

A group of children with dyslexia and a retained ATNR integrated the reflex by means of motor exercises. It turned out that the treated children improved their reading ability in a significant way compared to an equivalent control group that did not do any exercises to integrate the reflex and in whom it remained active.

A study from Belfast[99] of 739 children demonstrated that a retained ATNR deteriorated both reading and spelling ability in a significant way. The authors question that dyslexia is a limited phonological phenomenon and instead emphasize many studies that point to different causes.

Development of convergence and accommodation during infancy

During the first two months, the baby spends much of its time in the ATNR pattern, with her head turned in one direction. After that, the baby develops the symmetrotone reflex pattern, with her head and hands in the midline of the body, fixating her hands, playing with an object, or sucking her toes.[100] This pattern is suggestive of the second phase of the Moro reflex, when arms and legs are drawn to the midline of the body. Each time the Moro reflex is triggered, the baby will activate this symmetrical pattern. Normally the Moro reflex is integrated into the symmetrotone reflex pattern at the age of three or four months.

By means of this symmetrotone pattern, the infant will have an opportunity to practice focusing on an object at a near distance in the midline. In order for the picture to get to the fovea, or central visual field of both eyes, the baby must direct both eyes toward the object in the midline—that is, converge. In order for the image to be clear, he must accommodate—change the shape of the lens—which will help correct the inborn farsightedness of the child. By lying on his back, fixating his hands, grasping objects and moving them from one hand to the other, or putting his toes into his mouth, the baby will get basic training of both accommodation and convergence.

In a supine position, the eyes of the baby will focus at a near distance. When the baby lies on his stomach and starts to lift his head, the baby will get an overview of his surroundings and be able to focus on objects at a distance, and he may start crawling on his stomach to reach what he sees. When the baby learns to sit up and finally rises and walks, his focus will change more and more to objects at a long distance. Later, when the child learns to read and write, work at a near distance will become more important.

To be able to lift her head and later get up on hands and knees is crucial for the baby's ability to change focus from a near to a far distance, and vice versa, and that is the ability of accommodation. When the baby crawls on hands and knees, she can practice her ability to see at both short and long distances by looking down at the floor and forward. Before the baby is able to lift her head, the tonic labyrinth reflex must have been integrated at least to some extent. The Landau reflex must develop for the baby to be able to lift her head and chest from the floor. Before the baby is able to crawl on hands and knees, the symmetrical tonic neck reflex must have been sufficiently integrated.

To correct reading challenges due to visual challenges

Reading disabilities that are caused by visual factors can be remedied by different methods: visual correction with glasses, movement training and reflex integration, and exercises that improve visual skills. In reading disability, suspected to be caused by visuals factors, one should first recommend visual examination by an optician or oculist. Often this kind of reading disability improves considerably if proper glasses are prescribed. Long- sightedness that causes challenges with accommodation and binocular vision might easily be remedied by using proper reading glasses.

Regrettably, not all opticians and oculists are interested in making a thorough examination of binocular vision but limit themselves to examining whether there is an error of refraction. If there is only slight farsightedness and the optician or oculist does not believe in the connection between visual challenges and reading disability, he might think that the refraction error is too small to require correction by reading glasses and the student will not get necessary assistance. The child may also be given glasses that do not help his challenges, and therefore he might not want to use them.

Even worse, if the child, due to visual stress, has developed pseudo-myopia caused by ciliary muscle spasm, the child's vision could be seriously harmed if she gets a prescription for concave minus glasses for myopia. Such glasses would aggravate the ciliary muscle spasm instead of relaxing it, which should be done by means of concave reading glasses. Such mistakes are not infrequent.

It is therefore safer to propose that the child be examined by a behavioral optometrist, who not only regards possible refraction errors but also makes a thorough examination of all visual factors that may cause reading difficulties and makes a comprehensive assessment of the child and his situation.

Besides recommending a visual examination, the child should also be given the opportunity to improve her vision by rhythmic movement training and reflex integration. By means of such training, challenges with accommodation, pursuit eye movements, and binocular vision will improve. Another advantage of movement training and reflex integration is that it affects dyslexia not only by improving vision but in many other ways also.

Visual skills like accommodation, binocular vision, and saccadic eye movements will improve with a combination of visual and rhythmic exercises and reflex integration.

Checking of primitive reflexes in visual challenges

Based on a thorough visual examination, an experienced optometrist is able to conclude which active primitive reflexes are affecting vision. Many children with an active fear paralysis reflex and Moro reflex have an esophoria, which can be marked at both a close and long distance. If the child has esophoria only in short distance vision, which also improves if the child uses reading glasses, one can draw the conclusion that there is farsightedness and/or impaired ability of convergence due to a nonintegrated symmetrical tonic neck reflex.

Impaired ability of accommodation is a definite sign that the STNR has not been integrated. In farsightedness, impaired accommodation and an active STNR is usually seen. The STNR is usually active in myopia as well.

In children with strabismus, the asymmetrical tonic neck reflex (ATNR) is usually active, often more in one direction. The grasp reflex, the Babkin Palmomental reflex, and the hands pulling reflex are often active as well. In addition, difficulty with pursuit eye movements is caused by insufficient integration of the reflexes of the midline dimension, usually the ATNR.

If the visual examination shows that the child has challenges with esophoria or exophoria, accommodation, binocular vision, or so forth, it is important to make a thorough examination of the primitive reflexes and integrate the reflexes that are still active.

Reflex integration and motor training in visual challenges

Reflex integration and rhythmic movement training in combination with visual exercises are very effective methods in visual challenges. In many cases, it is possible to remedy challenges with accommodation and binocular vision quite fast by integrating the STNR and the ATNR reflexes with rhythmic exercises, in combination with visual exercises. This is especially suitable for children below the age of eight or ten. For older children and adults, a combination of isometric reflex integration according to the Masgutova method and visual exercises can be used. It is also very important to integrate the Moro and fear paralysis reflexes not only to correct esophoria at a far and close distance but also to facilitate accommodation by reducing inner stress and by reducing light sensitivity that may obstruct accommodation and cause the text to become blurry.

Case report

Pelle was eleven and had just learned to read when he started rhythmic movement training. He used to practice reading aloud with his mother. He would read one sentence and she the next one. He read with great effort in a book with big letters. When reading, he almost immediately lost his concentration, he soon got a headache, his eyes got tired and irritated, and the text started to jump. Otherwise, Pelle had no problems with attention and concentration. A few weeks after he started the training, he was diagnosed with dyslexia.

I tested his reflexes, and he had ATNR, STNR, Moro, and fear paralysis reflexes retained.

A visual examination with the Bernelloscop showed that he had esophoria around ten, both at a close and far distance. With reading glasses +1, his esophoria at close distance normalized.

He had a rotated pelvis, which was corrected when I integrated the STNR.

He was given homework to do: rhythmic exercises and isometric integration starting with the ATNR. Pelle did his exercises every day, and after two months, he could read for an hour without a headache.

Pelle continued his exercises and started to work with the STNR integration at home, assisted by his mother. After three months, a visual examination showed that his esophoria, both at a far and a close distance, had vanished.

After he had done the exercises for seven months, the STNR, ATNR, Moro, and fear paralysis reflexes were integrated. His reading had improved, and he could read more and faster than before. He still had problems with writing and spelling and started to work on hand reflexes: grasp, hands pulling, and Babkin reflexes.

One year after he started the rhythmic movement training, he loved reading, devoured books, and had started to borrow adult books from the library.

He now only had slight spelling problems, and his writing had improved considerably. He never had headaches or symptoms from his eyes, even when he read for long periods.

CHAPTER 16

PHONOLOGICAL AND WRITING CHALLENGES

Dyslexia and phonological challenges

According to dyslexia research, all dyslectics have phonological challenges. They have impaired auditory perception—that is, challenges in perceiving sounds. They have diffuse ideas about the sound structure of words, poor short-term memory for words, indistinct articulation, poor short time memory for sounds, difficulties in storing new names and words, and repeating complicated words.

Two factors may cause phonological challenges: impaired hearing and poor articulation. Both hearing and articulation can be improved by different methods. Hearing can be improved by auditory stimulation, and motor training and reflex integration can improve articulation.

Ear dominance and dyslexia

Sounds that we hear with our right ear predominantly go to our left auditory cortex, and vice versa. In about 90 percent of all people, the language center is situated in the left hemisphere. In most people, the right ear is the dominant one, which is practical since, in this case, most sounds of the language primarily go directly to the speech areas of Wernicke and Broca, in the left hemisphere.

Danish psychologist and dyslexia researcher Kjeld Johansen[101] has shown a connection between dyslexia and dominance of the left ear. According to an investigation he made, more than half of a group of dyslectics were dominant with the left ear. Left ear dominance usually makes it more difficult to perceive spoken language since sounds from the left ear must make a detour to the right hemisphere before they reach the language areas in the left auditory cortex, where they are processed. This applies to the 90 percent of the population who have their language centers in the left hemisphere. In such cases, the hearing process will be delayed and the individual will have difficulties catching on to what is being said.

Another investigation, which Kjeld Johansen made, showed that many dyslectic children had suffered from recurrent ear infections when they were babies. Due to impaired hearing caused by these ear infections in a critical period of the development of hearing, the ability to discriminate between different sounds suffered because of the more or less temporary insufficient stimulation of the auditory cortex. Therefore, this decreased ability to discriminate between different sounds of language will persist even after the ear infection has healed. The investigation showed that right ear inflammation is more serious since it makes it more difficult for the right ear to become dominant.

Phonological challenges and the cerebellum

Phonological ability is not only dependent on hearing. Russian scientist Alexander Luria has called attention to the fact that in disorders of the sensory cortex in the left parietal lobe, the sounds *b* and *m*, which are pronounced with the same lip movements, are easily mixed up. Moreover, individuals with such disorders also have difficulties discriminating between the sounds *d*, *e*, *n*, and *l*. This indicates that the phonetic analysis is not only dependent on hearing but also on articulation.[102]

Research has shown that challenges with the cerebellum can cause speech difficulties. Individuals with lesions of the cerebellum often have poor articulation.[103]

The most important cause of poor articulation is malfunction of the cerebellum. This can be seen in children with severe motor challenges (for instance, in cerebral palsy, in which the cerebellum has not been properly developed due to insufficient stimulation). In these cases, articulation will improve when the motor abilities improve. Malfunction of the cerebellum can also be caused by inflammation of the cerebellum due to gluten sensitivity; casein intolerance; heavy metal toxicity, which is common in autism; and late speech development. In these cases, the articulation will improve with a proper diet in combination with rhythmic exercises. If the inflammation of the cerebellum can be limited by diet and detoxifications of heavy metals, the rhythmic exercises will stimulate the nerve nets of the cerebellum and articulation will get better.

In my experience, most children with late speech development have challenges doing the rhythmic exercises in a coordinated and rhythmic way, indicating a malfunction and inflammation of the cerebellum, usually due to gluten sensitivity and casein intolerance. Provided they are given gluten- and casein-free diets in order to diminish the inflammation of the cerebellum, speech will improve concurrently with the ability to do the rhythmic exercises. The more time it takes to learn to do the exercise rhythmically, the slower speech development will be.

Phonological challenges and the Babkin reflex

Logically, our articulation is dependent on our ability to control fine motor ability of the lips and tongue. Less obvious, maybe, is that it is closely connected to the fine motor ability of our hands. But there are many indications that movements of the mouth and hands are connected. Dyslectics often move their lips and hands when they write or cut with a pair of scissors or do various fine motor precision work. A similar pattern can be found in newborn infants. In infants, movements of the mouth can also trigger movements of the hands. When they are sucking the breast or the bottle, they can be seen opening and clasping their hands. At a later stage, when they begin to investigate their surroundings, the connection between hand movements and mouth movements is of great importance. When the baby grasps an object with his hands, he begins to make sucking movements with his mouth, and then he examines the object by putting it into his mouth. Signals from the tactile sense and the kinesthetic sense of the mouth and hand will stimulate the sensory cortex of the parietal lobe.

It is no coincidence that the nerve connections from the hands and the mouth/tongue region go to two areas of the cortex that are very close to each other and take up a considerable part of the sensory cortex. The connection between movements of hands and mouth is due to the Babkin reflex. This reflex can be triggered by a light pressure in the palms of the infant, who will open his mouth and bend his head forward or to the side.

The Babkin reflex should be integrated at the age of four months, when the baby has learned to grasp objects and to put them in her mouth. If the reflex is not integrated, the voluntary fine motor control of hands and mouth will be impaired. It may also cause tension in the jaw, grinding of the teeth, and challenges with articulation. An active Babkin reflex may also cause poor motor ability of hands and fingers, challenges in doing up buttons and tying shoelaces, and poor handwriting. Low muscle tone and over flexibility of hands and fingers are common signs of an active Babkin reflex. Involuntary movements of one's mouth when writing or playing an instrument is a reliable sign that the reflex has not been integrated.

Due to the difficulties of articulation, the corresponding area of the sensory cortex of the left parietal lobe is not properly stimulated, which can explain impaired phonological ability and difficulties in perceiving sounds. With motor training and integration of the Babkin reflex, articulation will improve, as will phonological ability.

Writing challenges

Writing difficulties are not only caused by retention of the Babkin Palmomental reflex. Other primitive reflexes are usually also involved, primarily the grasp reflex and the hands pulling reflex. A week after delivery, the grasp reflex can normally be triggered by putting a finger in the palms of the baby, who will grasp the finger and hold it in a tight grip even if you lift up the baby. The arms of the baby will be straight. The grasp reflex should be integrated by the age of one year.

The hands pulling reflex is normally active immediately after delivery. It is triggered by holding the baby around his wrists and pulling him toward you. Then the baby bends his arms and helps to get up into a sitting position.

At the age of two months, the grasp reflex is merged with the hands pulling reflex and they start to function as a unit: when you put your fingers in the palms of the baby, she will clutch your fingers and bend her arms so you can help her get up into a sitting position. These two reflexes are important for the ability of the baby to grasp objects and put them into her mouth, therefore assisting the integration of the Babkin reflex.

If the grasp reflex is not integrated, fine motor ability is affected and the child will hold the pen tightly, write laboriously, and have poor handwriting. An active grasp reflex causes tension in the shoulders, and an active hands pulling reflex causes tension in the forearms, making writing difficult. An active asymmetric tonic neck reflex will also obstruct writing because of tension in shoulders, arms, and fingers. If one or more of these reflexes is active, it will be difficult for the child to achieve automatic writing, and the child will have to focus consciously on the laborious writing process instead of focusing on what he wants to express. Such children can have difficulties writing essays because their consciousness is too focused on their writing.

Improvement of visual and articulation problems in cerebral palsy

Children with cerebral palsy often suffer from strabismus or refraction disorders. They frequently also have speech challenges. If they start talking at all, their articulation is usually poor. Such problems are seldom caused directly by a brain lesion but are rather a consequence of the impaired motor development due to the cerebral palsy. As stated above, the nerve nets of the cerebellum do not develop properly due to insufficient motor stimulation. In small children with such problems, speech and visual abilities usually improve considerably when their motor abilities are developed by rhythmic movement training. Two short case studies may illustrate this.

Lisa was four years when she started rhythmic movement training. She was hemiplegic (i.e., one side was more paralyzed than the other). She could speak a few words, but her pronunciation was such that only her parents understood what she said. She had a severe strabismus of one eye. She could not sit on her own, and lying down she would have her arms bent to her chest and her hands clutched.

A few weeks after starting the rhythmic exercises, her arms could be relaxed and she started to grasp things with her hands. After a few months, she learned to sit on her own, supporting herself with one hand. By then, her strabismus had improved considerably, and after a year, it could only be observed when she was extremely tired. She learned how to draw and paint and loved to do both.

Her speech developed considerably, as did her articulation. After four months, she could tell long fantasy stories about princes and princesses. After a year, it was no longer difficult to understand what she said, although her articulation was not perfect.

Eva, whom I have written about above, was three when she started rhythmic movement training. She could not sit on her own without being supported, and she could not use her hands. She had not started to talk, and the doctors had concluded that she would never talk and that she should start learning sign language. She was farsighted and had strong glasses.

Not long after she started the training, she learned to use her hands and to eat and drink on her own. After a few months, she began to speak, first occasional words and then two-word sentences, and after a year, she was saying sentences of up to six words. By then, she had also learned to do a jigsaw puzzle and dress and undress her doll. Her vision had improved considerably, and she had to have much weaker glasses.

How the training improves vision and articulation

Thanks to the rhythmic training, both Eva and Lisa were able to start using their hands.

When Eva, after her first visit to Kerstin Linde, learned to sit on her own without being supported, she could start to use her hands for the first time and create her own world. Her focus moved from a far distance to a short distance, and she had to accommodate more. By rocking on hands and knees and crawling, her symmetrical tonic neck reflex was integrated, which improved her accommodation and decreased her farsightedness. Her language developed simultaneously with her motor abilities due to the stimulation and development of the nerve nets in the cerebellum.

For Lisa, it was much more challenging to learn to sit without being supported, and therefore it took more time for her to learn how to draw and write. When her gross and fine motor ability improved, her articulation became more clear due to both stimulation of the cerebellum and improvement of her fine motor ability. At the same time, her strabismus started to disappear. After a year, her pronunciation was quite understandable and her strabismus was all but gone.

Case study

Repeatedly I am amazed by the effectiveness of rhythmic movement training in late speech development. In my work as a school doctor, I met a girl, Hanna, who was in second form. Her speech development was such that she had been placed in a class for children with special needs. She did not want to speak with me, but her teacher had informed me that she never seemed to understand the topic of a conversation and her answers would be constantly out of place. In addition, her grammar and syntax made her speech difficult to understand. She had been diagnosed with late speech development and autistic spectrum disorder with a question mark.

I showed her three simple rhythmic exercises—the windscreen wipers, sliding on her back, and rolling her bottom from side to side. The latter she did with some difficulty. I told her mother that Hanna should do these exercises every day in order to improve her speech.

After a year, I met Hanna again. Now she spoke fluently and comprehensibly with me and answered my questions adequately, with correct grammar and syntax. Her mother told me that Hanna had done the exercises every day for eight months, during which her speech had continuously improved. Then she had made a break and had not resumed the exercises again.

Her teacher confirmed that her speech problems had petered out and now were gone, and concurrently so were many of her autistic symptoms. However, she had not yet learned to read, indicating that she would probably need to work on more than her speech challenges.

The importance of the cerebellum in dyslexia

Many children with dyslexia have challenges doing the rhythmic exercises in a coordinated and rhythmic way, which indicates a dysfunction of the cerebellum. Such a dysfunction may cause challenges with reading in different ways. As explained above, it may cause articulation challenges and phonological problems, which will improve when the cerebellum is stimulated by the rhythmic exercises.

Another function that is important for reading is eye movements. They are controlled from an area in the frontal lobe that has important nerve connections with the cerebellum. In some children, difficulties with saccadic eye movements are connected with a dysfunction of the cerebellum. In such cases, rhythmic exercises in combination with a visual exercise for saccadic eye movements can improve speed of reading to a great degree.

The cerebellum also has a great importance for the function of the prefrontal cortex, which plays a key role in learning how to read and reading comprehension. The rhythmic exercises will stimulate the prefrontal cortex by way of the cerebellum, thus making it easier for children to learn how to read and to understand what they are reading.

CHAPTER 17

THE NEURAL NETWORK OF READING AND THE PREFRONTAL CORTEX

The neural network of reading

As has been shown, we must use and coordinate different senses and abilities in order to read. The visual, auditory, and proprioceptive senses all contribute to the decoding of the text. If the sensory processes do not function adequately, corresponding areas of the brain do not develop properly. If we had challenges in hearing as infants, the auditory cortex may not have learned to discriminate different sounds from each other. If vision does not function properly, the visual cortex might not develop properly. Challenges with the fine motor ability might cause challenges with articulation, and challenges with articulation might cause malfunction of the sensory cortex of our parietal lobes. If the cerebellum does not get adequate stimulation due to an inability to perform rhythmic movements, there might be lack of stimulation of the frontal lobes and especially areas of the left frontal lobe that are essential for spoken language. If our body halves cannot cooperate due to motor challenges, the nerve fibers through the corpus callosum that connect the brain hemispheres will not get sufficient stimulation. There will be insufficient myelination of these fibers, obstructing the cooperation of the body halves and eyes.

By sound stimulation, the auditory cortex can be stimulated and hearing improved; and by prescribing glasses, refraction errors can be taken care of, thereby improving vision. This can help improve reading ability. But as has been shown by motor training vision, articulation, phonological ability, cooperation between the hemispheres, eye movements, and many other things can be affected through stimulation of those areas of the neocortex that are responsible for these abilities.

By measuring the blood flow of the brain with PET, it has been shown that several areas of the neocortex are activated when we read, referred to as the neural network of reading.[104] The visual cortex of the occipital lobes, areas of the temporal lobes and the frontal lobes that are in charge of grammar and phonetic analysis, areas of the motor cortex of the frontal lobes that control motor speech ability, and areas of the frontal lobes that control eye movements are all essential for the neural network of reading. Each of these areas must be sufficiently developed, and they must be sufficiently linked up with each other both in the same hemisphere and in the opposite one in order for reading to function without challenges.

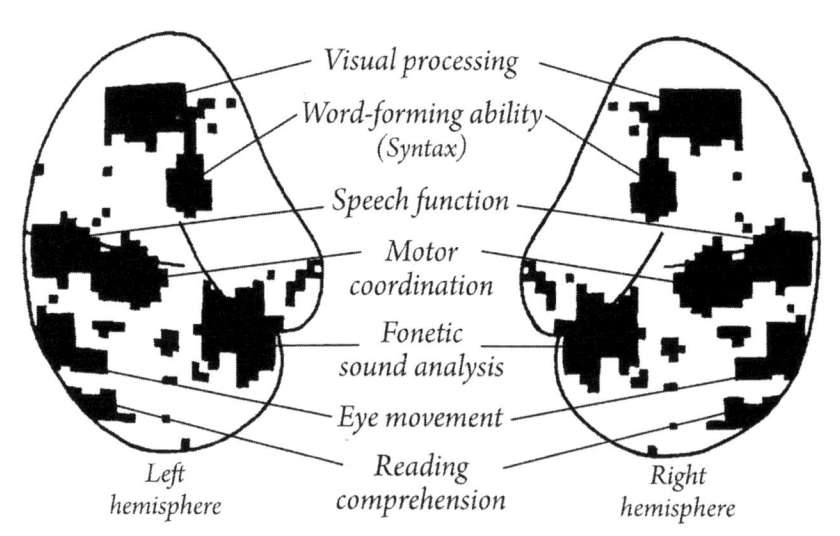

Fig. 27 The neural network of reading

If the corpus callosum does not operate properly, the exchange of information between the hemispheres will be impaired. In spite of the fact that language and speech centers are situated in the left hemisphere, blood flow measuring shows that both hemispheres are equally activated when we read.

Since the areas of the neocortex that are activated during reading are situated far from each other, the transmission time of nerve signals are of crucial importance. If the transmission of signals between the areas that are involved is slow, reading will be impaired. Speed of transmission is dependent on the myelination of the nerve fibers that are involved. Myelination is stimulated when a nerve path is being used. Transmission time is therefore dependent on the frequency of utilization of the nerve paths between the different areas of the neural network of reading. The more these nerve paths are used, the quicker the transmission. Thus the more a child reads, the more it is stimulated.

Motor training also stimulates myelination. Various kinds of cross movements stimulate the myelination of the corpus callosum, and exercises that develop fine motor abilities, eye-hand coordination, and so forth, stimulate the myelination of the nerve pathways between different centers that are part of the neural network of reading.

The prefrontal cortex—the chief executive officer of the brain

When we read, areas of the neocortex that process information received from our senses are activated. But according to blood flow measuring there is another area, whose function for a long time was not well understood, which plays a major role in reading. That is the prefrontal cortex, which is situated at the very front of the brain. Elkhonon Goldberg calls this the chief executive officer of the brain.

"The prefrontal cortex plays the central role in forming goals and objectives and then in devising plans of action required to attain these goals. It selects the cognitive skills required to implement the plans, coordinates these skills, and applies them in a correct order. Finally, the prefrontal cortex is responsible for evaluating our actions as success and failure relative to our intentions."[105]

When we learn how to read, the task of the prefrontal cortex is to mobilize and direct the neural network of reading and the various areas that take part in the reading process. If these areas are not developed enough due to insufficient stimulation or if the lines of communication do not work properly due to insufficient myelination, the prefrontal cortex cannot fulfill its task of directing the learning process and the child might develop a reading disability.

Reading comprehension and the prefrontal cortex

Once we have learned how to read, the prefrontal cortex no longer plays a central role in the reading process, but in order for us to comprehend what we read, the prefrontal cortex must work adequately. Reading comprehension is dependent on our ability to create something in our consciousness that cannot be found in the text—let us call it our personal home video.

When we read cookbooks with recipes of our favorite dishes, the text calls to mind images as well as taste and smell sensations that we evoke from our memories. Our reading comprehension is largely dependent on our experiences, which are stored as memories in various places of the brain. The task of the prefrontal cortex is to produce memories that correspond to the text or to our intentions and make them available to our consciousness in the reading process as well as in real life. When we want to wear our shoes, we know where to look for them if we have a memory of where we put them, and we engage the prefrontal cortex to make that information available. If the prefrontal cortex does not function as it should, we may not know where to look for our shoes, and we might even end up putting on a cap instead of a pair of shoes.

It is the same thing when we read. If the prefrontal cortex does not work properly, we will not be able to play the relevant home video corresponding to the text, and when we read our cookbooks, we will not see the images and feel the smell and taste of our favorite dishes.

Dyslexia and the prefrontal cortex

In dyslexia, the CEO of the brain, or the prefrontal cortex, does not work properly, impairing reading, learning, and comprehension. There may be various reasons for the dysfunction of the prefrontal cortex. For example, the prefrontal cortex may get insufficient stimulation from the senses via the reticular activation system (RAS) or too little stimulation from the cerebellum due to a dysfunction of the cerebellum. There also may be insufficient stimulation from the areas of the brain that are part of the neural network of reading.

Lesions of the prefrontal cortex affect the function of the whole brain. In reverse, lesions or dysfunctions in various places of the brain will have repercussions on the prefrontal cortex. This is easy to understand if the relationship between the prefrontal cortex and the rest of the brain is compared to the relationship between a commander and his army. If the commander is wounded, the lower units will get no proper guidance and chaos may follow. If, on the other hand, the lower units are disorganized or the lines of communication do not work, the ability of the commander to give relevant orders will be obstructed.

Research has shown that the blood flow of the frontal lobes will decrease if there are lesions in other areas of the brain.[106] When there is a dysfunction of different areas of the cortex that are part of the neural network of reading or the neural connections between these areas, there will be repercussions for the performance of the prefrontal lobes. As we have seen, inabilities may be found in dyslectics that mirror dysfunctions in different areas of the brain. Phonological challenges may be due to dysfunction of the auditory cortex of the temporal lobe, the sensory cortex of the parietal lobe, and the motor cortex of the frontal lobes. Besides, the lines of communications between these areas might be insufficient. All these dysfunctions impair the performance of the frontal lobes.

Likewise, challenges with vision and eye movements can be signs of dysfunctions of the visual cortex or the area of the frontal lobes that control eye movements, dysfunctions that also impair the function of the prefrontal cortex.

Dyslexia and challenges with attention

Many dyslectic children also have challenges with attention and concentration. One common cause of such challenges is lack of stimulation of the cortex and especially the frontal lobes from the senses via the reticular activation system of the brain stem. Another important cause is insufficient stimulation of the frontal lobes from the cerebellum. Increased distractibility due to retained fear paralysis and Moro reflexes are also very common causes of challenges with attention.

Signals from the tactile, proprioceptive, and vestibular senses are primarily responsible for the stimulation of the neocortex. Children with motor handicaps; poor muscle tone; and primitive reflexes that are still active, especially those that affect the ability to focus and that cause a shrunken-up body posture, get insufficient stimulation of the frontal lobes via the reticular activation system.

Children who have not been able to make rhythmic baby movements because of challenges with the cerebellum or other motor challenges will not activate the neocortex, especially the frontal lobes, sufficiently.

Once children with attention challenges have learned to read, they need not have great challenges in decoding the words if their neural network of reading is functioning reasonably well. They may even be fluent readers. This is especially true if the ATNR is integrated. But they often have problems understanding what they read. The reason for that is that the neocortex, especially the prefrontal cortex, is insufficiently activated. This means that the prefrontal cortex cannot fulfill its task to direct the personal home video. It is as if the child cannot switch on his video. He reads, but nothing happens. Even if he reads perfectly, he has problems understanding what he reads.

Rhythmic movement training and the prefrontal cortex

If the prefrontal cortex does not work properly, the child may have challenges learning how to read. Once reading has become an automatic process, the child will be able to read even if the frontal lobes are damaged; however, reading comprehension will suffer. It is therefore important to help children with reading challenges to improve the function of the prefrontal cortex. This can be done in different ways. To begin with, by means of rhythmic movement training, muscle tone and posture can be improved in order to bring about more stimulation of the neocortex and the prefrontal cortex by the reticular activation system. Additionally, nerve paths from the cerebellum can stimulate the prefrontal cortex by the rhythmic exercises. Finally, we can stimulate different areas of the neocortex that are part of the neural network of reading and consequently stimulate the prefrontal cortex. When the visual skills, articulation, fine motor abilities, and so forth, are trained and developed by motor training and reflex integration, as has been described above, the ability of the prefrontal cortex to act as the chief executive officer of the brain is also improved.

Different causes of dyslexia demand different approaches

Some people with reading challenges are easy to help. Individuals with poor reading comprehension who read fluently will benefit rapidly from doing rhythmic exercises and integrating primitive reflexes. Most commonly, they need to work on the TLR, STNR, and often the fear paralysis and Moro reflexes. In such cases the ATNR and other midline reflexes are usually adequately integrated, which also explains why such individuals usually had no problems learning how to read.

Children without attention problems who have problems learning to read because of visual challenges may start reading rapidly as soon as their vision improves. In such cases, the priority would be to work with the visual challenges by integrating the ATNR, STNR, fear paralysis, and Moro reflexes. This can be done both by rhythmic exercises and isometric integration exercises, as illustrated by the following case study.

Case study

A twelve-year-old girl had never been able to read more than one or two sentences because the text was always jumping and becoming blurry when she was reading, making her completely exhausted. She had no severe attention problems and from the start could take part in the isometric exercises without any problems. Doing the ATNR integration exercise, focusing at a point with her eyes, she experienced double vision and some irritation and tension of her eyes. She was asked to do rhythmic exercises every day and the isometric integration of the ATNR three times a week. After a couple of months, her reading improved somewhat as the visual symptoms during reflex integration improved. She was now able to read a page before her eyes got tired, the text got blurry, and the text started to jump. She now reported that this would happen sooner when she was looking at a text the teacher had written on the board.

When I tested her for the STNR, she turned out to have a rotated pelvis, and when she did the isometric integration for the STNR exercise, the point she was looking at started to move and her vision became blurry, indicating poor stability of accommodation. Before the isometric integration was finished, the point had stopped moving. Also, her pelvis was no longer rotated after integration. She was told to go on doing rhythmic movements every day and the isometric exercises for the STNR and ATNR, each twice a week. When I saw her a few months later, her reading had improved dramatically. During summer vacation, she had been able to read three books without any visual problems.

Attention problems and dyslexia

When a child with dyslexia has many challenges, it will take more time and effort for him to overcome his reading difficulties. Attention problems and poor endurance often need to be dealt with before the child is able to improve visual challenges with isometric exercises. Sometimes there may be severe challenges doing the simple rhythmic exercises, reflecting a dysfunction of the cerebellum that may cause problems with language, articulation, eye movements, or the functioning of the prefrontal cortex. In these cases, stimulating the cerebellum with rhythmic exercises often needs to be prioritized before working with reflexes and visual challenges. Sometimes food intolerance may trigger the stress reflexes and obstruct concentration and reading comprehension. However, every child is different and needs an individual approach.

The more challenges a child has, the more time and effort it will usually take before reading is fluent. However, this is not always the case, as the flowing case study may illustrate.

Case study: Maria

The case of Maria can illustrate how motor training can stimulate the neural network of reading and the frontal lobes and what the result can be.

Maria was twelve years when she started rhythmic movement training. Her condition resembled slight cerebral palsy, although she had never received that diagnosis. Her hips were very much rotated inward, and she stumbled on her feet when she tried to run. She did not lift her feet when walking, and she had a shuffling gait. Her articulation was poor, and she talked very indistinctly. She regularly visited a speech therapist. She had poor fine motor ability and had not developed a correct pen grip. Her posture was poor, with her back hunched; her arms were weak, and she was not able to keep her head in an upright position most of the time. Her balance was poor. She found reading difficult and could in fact only sound out single words. She experienced that the text jumped around on the page. She had been prescribed reading glasses but did not like to use them.

A visual examination with an instrument showed marked esophoria at both close and long distances. There was no fusion, and sometimes she suspended the use of one eye. When she used her glasses, her esophoria at a close distance was almost normalized and there was fusion. Her pursuit eye movements were poor.

Maria had many active primitive reflexes. All reflexes related to cooperation between the hemispheres were active: the ATNR, the grasp reflex, the Babkin reflex, the hands pulling reflex, the Babinski reflex, and the leg cross flexion reflex. The two latter reflexes were strongly active, and they may explain her inability to run. Her articulation challenges may be explained by the Babkin reflex and dysfunction of the cerebellum due to her motor handicap. The grasp reflex explained her poor fine motor ability and absence of correct pencil grip.

Affecting Maria's ability to accommodate was the retention of the STNR. The STNR and the TLR explain her hunched posture, weak arms, and inability to keep her head in an upright position. The TLR explains her poor balance, and the STNR her farsightedness and tendencies of suspension, making the text jump when she did not use her reading glasses. She also had an active Moro reflex that contributed to her esophoria.

Maria started rhythmic movement training and reflex integration. She did not get a lot of support from her parents, though her grandmother used to periodically help her. Besides rhythmic movements, great emphasis was placed on isometric integration of her primitive reflexes, which her grandmother helped her with. After a few months, her reading improved. After half a year of training, she learned how to run. Her balance improved. She straightened her back and was able to keep her head in an upright position. Her articulation improved considerably.

After a year of training, she could read quite well without reading glasses. An examination of her vision showed that her esophoria had improved, especially at a close distance, where she had good fusion without tendencies of suspension. She began to play basketball and became a good player, and she had no problems running.

Rhythmic movement training instead of remedial teaching in a Swedish school

In one Swedish school, there were challenges a few months before the third-grade students were to move to grade four. Nine of the students were such poor readers that their teacher estimated that the school needed to employ a teacher part time for remedial teaching. The remedial teacher tested grade two, and the result showed that their reading was so poor that they needed remedial teaching at that level. Instead of hiring a part-time teacher, it was decided to try rhythmic movement training and reflex integration. In February, the group started reflex and rhythmic movement training once a week. Lars-Eric Berg tested all the students and gave them individual training programs. The remedial teacher was present and learned how to work with motor training during her classes. The parents were instructed to help their children with rhythmic exercises every day. No additional remedial education for reading was started.

After three months, the remedial teacher evaluated the reading ability of the children. All students, except one, were normal readers for grade three. In three months, the group had made up for one year of reading development *by using only motor training.*

The parents also reported many positive side effects of the training. The motor ability of some of the boys had improved to such a degree that they now qualified to play on the football team, where they rarely had been admitted before.

One girl who had kept to herself and had no friends started to invite other children home and joined the Girl Scouts.[107]

Contradictory expert opinion on motor training in dyslexia

According to Swedish dyslexia experts, as stated in the consensus report by twenty-two dyslexia researchers, the cause of dyslexia has now been conclusively and scientifically confirmed. Dyslexia is a phonological problem and nothing else. It should be treated with intensified reading practice and better reading instruction. Any positive effects of motor training on dyslexia are flatly denied. The experts even claim that all efforts that focus on anything other than the linguistic development of the student will have negative effects "since they reduce the scope for efforts that by evidence will lead to improved reading and writing development."

However, if one looks beyond the backwater of Swedish dyslexia research, one will find that dyslexia researchers do not share this conviction internationally. Many of the causes of dyslexia that I have discussed in this book have been objects of studies in international dyslexia research.

In their article cited above, M. McPhilips and J. Jordan-Black refer to a number of articles published during the last years, showing that children with dyslexia "have problems that extend beyond the range of underlying language-related deficits." A number of studies have shown that some children with dyslexia have impairment in several areas of the visual system (Stein and Walsh 1997) and that auditory temporal processing may be impaired (Witton el al. 1998). Furthermore, dyslexic performance is often characterized by poor motor skill and poor balance, and it has been suggested that a dysfunction of the cerebellum might underlie the major deficits seen in dyslectic children (Fawcett, Nicolson, and Dean 1996). Neuroimaging studies have shown abnormal activation and morphology of a number of brain regions in adults with dyslexia, including the cerebellum (Rae et al. 2002). Finally, there is considerable evidence of comorbidities involving reading difficulties, attention deficit, and motor coordination deficits" (e.g., Iversen, Berg, et al. 2005; Visser 2003).[108]

All these findings have caused scientists to question that dyslexia is a limited phonological phenomenon. According to the researchers behind the study from Belfast and the quotation above, these results support an understanding that is becoming more widespread so that "the development of literacy is dependent on a complex interaction of cognitive, environmental, and biological factors over time."

APPENDIX

Rhythmic Exercises

Rhythmic movements in a lying position

Rhythmic movements for stimulating the nerve chassis, or brain stem

The fetus gets sensory stimulation from the mother's respiration, heartbeat, walking, running, and so on. Such passive stimulation affects the tactile, proprioceptive, and vestibular senses of the fetus and stimulates the growth and maturing of the nerve cells of the brain stem. All such stimuli also promote the maturing of other parts of the brain.

Also, when primitive reflexes are triggered in the fetus, the motor responses cause stimulation of the nerve chassis. Another source of sensory stimulation for the fetus is all movements it can do by itself, such as turning his head from side to side, sucking his thumb, playing with its umbilical cord, and so on. Playing with the umbilical cord will cause proprioceptive stimulation, which has a calming effect on the fetus.

Passive Rhythmic exercises are especially useful in stimulating the brain stem in infants as well as in children with brain injuries who are still neurologically on the level of infants. By passively rocking the baby in different ways, the brain stem will be stimulated to improve muscle tone. Such rocking also helps to mature primitive reflexes and stimulate spontaneous infant movements. When an infant is slow to develop and does not easily progress from one stage of development to another, such stimulation can be used to quicken development (e.g., with children who are not able to lift their heads or do not start crawling on hands and knees). These exercises are also helpful in improving muscle tone in floppy children who are not able to move sufficiently on their own. In a child with severe brain damage who is stuck with his head turned to one side, passive rhythmic stimulation may cause him to turn his head from side to side in a reflex-like manner.

The following are movements that can be used for stimulation of the brain stem in infants, children, and adults.

Movements for the Nerve Chassis

1. Passive rhythmic stimulation from the feet in a supine position

The client lies on his back with his arms along his sides. Notice if the head is straight or is bent to one side. Notice the position of the feet. Normally a client should lie symmetrically, with an angle of forty-five degrees, to the floor.

Take a steady grip around the balls of the feet of the client and rock his body in the longitudinal direction. Try different speeds and notice when the client falls into a smooth rocking rhythm. The movement should be smooth and easy, not jerky and difficult.

Notice tensions on any levels that prevent the flow of the movement. If the ankle joints are tense, it may be easier to take hold of the ankles instead of the balls of the feet. Often the flow stops at the level of the diaphragm or the neck, sometimes the hips.

If the client does not lie straight but bends his neck to one side, an active asymmetric tonic neck reflex (ATNR) may be suspected. A stiff neck that does not follow the movement indicates one or more active neck reflexes: ATNR, TLR, fear paralysis reflex, or STNR. Persistent folding of the arms across the chest instead of keeping them stretched along the body indicates an active Moro reflex.

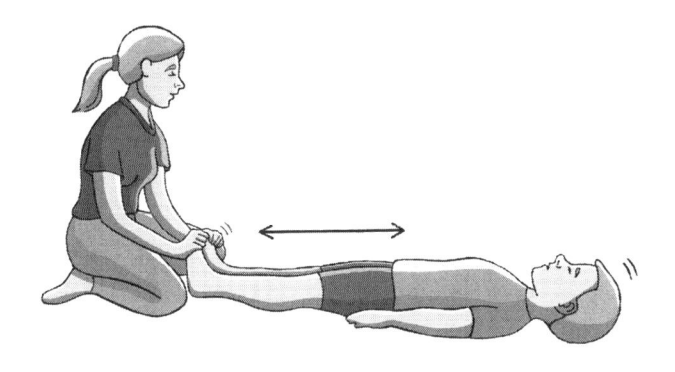

Fig. 28 Passive rocking from the feet

Please do not do this movement in Down syndrome if there is a malformation of the neck vertebrae!

2. Passive stimulation from the knees in supine position

The client lies on her back with her arms stretched along his sides and her knees bent fifty to sixty degrees. Hold under the knees and push rhythmically toward the head. If it is difficult to obtain a flowing rhythmic movement by pushing, you can instead pull rhythmically either by holding just above the knees or on the calves, just below the knees. Notice stops in the flow at different levels. This movement helps to integrate the spinal reflexes.

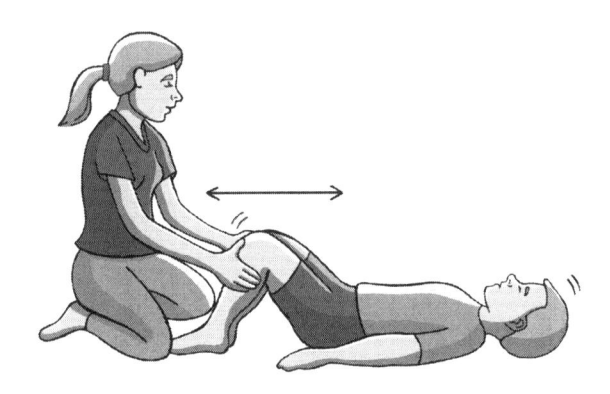

Fig. 29 Passive rocking from the knees

Please do not do this movement in Down syndrome if there is a malformation of the neck vertebrae!

3. Passive stimulation from the hip in a fetal position

The fetal position for passive rhythmical stimulation is excellent because the flow seldom stops at any level. Most efficient is to rock along the spine from the hip toward the head. In that way, the whole back and the head are involved in the movement. Put one hand on the seat bone and rock along the spine. If there is instability in the waist region and the upper body is waving from side to side, put the other hand on the shoulder for stabilization.

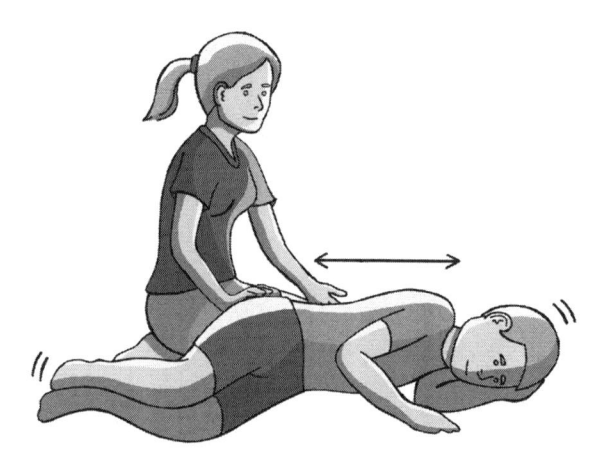

Fig.30 Passive rocking in a fetal position

4. Passive stimulation from the chest

The client lies on his back with arms along his sides. Place your hand on one side of the rib cage and gently rock side to side. This movement provides a lot of stimulation of the intestines.

Alternatives to this movement are to come from the opposite side of the chest or to push/pull from both sides of the chest.

Fig. 31. Passive rocking from the chest

5. Passive rolling of the bottom in prone position

The client lies in prone position with her forehead on her hands. The armpits should be close to the mattress. Hold the waistband and roll the bottom gently from side to side. The movement should start at the level of the nipples. If the shoulders and head are moving, put one hand on the shoulder blades to stabilize. The ankles should be relaxed and resting on the mat. If not, put a pillow under them.

Fig. 32 Passive rocking of the bottom

6. Rotation of the head from one side to the other

The client lies on his back and rotates his head from side to side. This exercise can be done in different ways: slowly with large deflections or small quick deflections in the middle line. Check that the movement is made symmetrically and correct if necessary.

This exercise stimulates the vestibular sense and the proprioceptors of the neck and causes relaxation not only of the neck but also of the whole body.

Fig. 33 Rotation of the head

Please do not do this movement in Down syndrome if there is a malformation of the neck vertebrae!

Rhythmic movements for the cerebellum

The cerebellum in attention disorder

The strong impact on attention, concentration, control of impulses, abstract thinking, judgment, and learning that is usually seen in rhythmic movement training can be explained by various factors such as improved arousal of the cortex or increased stimulation of different areas of the cortex by the cerebellum. However, children who have difficulty doing rhythmic movements in a smooth, rhythmic way due to a dysfunction of the cerebellum do not benefit as much from the movements as children with no such difficulties.

It is therefore most important to teach these children to make the movements in a rhythmic way. Some of them learn quite quickly, within a month or so, but others may have to practice daily for more than a year before they can make the movements smoothly, rhythmically, and effortlessly. Even then, they tend to lose the rhythm when they are tired. Rhythmic movements that are made actively are most important for remedying any dysfunction of the cerebellum. In addition, these movements have other effects such as integrating primitive reflexes and developing postural lifelong reflexes, especially in small children. These movements also promote the linking up of different parts of the brain—for instance, by stimulating the growth of nerve nets that are essential for the arousal of the cortex and the stimulation of different areas of the cortex by the cerebellum. All these effects are most important for resolving attention and learning difficulties.

It takes time to rebuild the brain in this way; therefore, the brain needs continuous and daily stimulation from active rhythmic movements for a long time, most commonly a year or more before learning and attention problems can be completely resolved.

Movements for the cerebellum

The following are the most important movements used for the stimulation of the cerebellum.

7. Sliding on one's back

The client lies on his back in the same position as in movement number two above and makes the same movement by rhythmically pushing from his feet. This movement demands some degree of coordination and can be difficult for some clients, especially those with a dysfunction of the cerebellum or an active spinal galant or spinal Pereze reflex. In that case, you can help the client first by pulling rhythmically from his knees until the client can make the movement on his own or perhaps help with light rhythmic touches to his knees.

Notice if the flow stops in the neck. If so, you might ask the client to let his head follow the movement by nodding his head voluntarily. Notice if the client uses his shoulders or arms to assist the movement. If the client is doing this, gently point it out and assist the person in stopping this.

Stopping of the flow of the movement in the neck indicates active neck reflexes. Accessory movements in the shoulders or hands may indicate insufficient grounding and poor cooperation between upper and lower body due to nonintegrated reflexes.

This exercise is excellent for integrating the spinal galant and spinal Pereze reflexes. It gives vestibular stimulation, emotional relaxation, and activates the cortex, diminishing hyperactivity and improving attention.

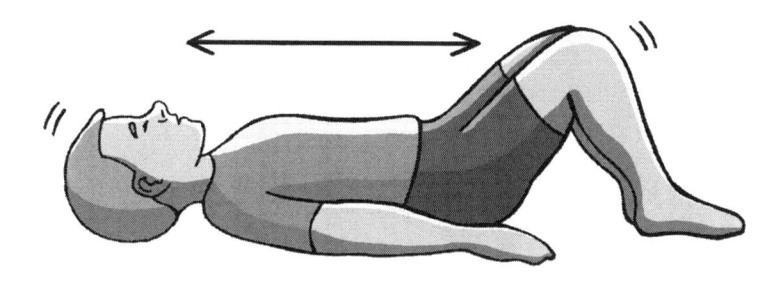

Fig.34 Sliding on the back

8. Rotation of the legs to make the big toes meet in the middle

The windscreen wiper

The client lies on his back with his feet about 10 cm (4 inches) apart. Notice the position of the feet. Are they symmetrical? Then ask the client to rotate his legs and make the big toes touch each other in the middle. Notice if the movement is rhythmical and symmetrical. The movement should be made with large deflections from the middle to the floor as far as possible without losing a spontaneous rhythm. The movement should start from the hips and the feet should not be involved. If the client cannot do this exercise without moving his feet it is a sign that the Babinski reflex is active.

Many children have challenges to do this movement rhythmically which is a sign of a dysfunction of the cerebellum. Adults usually have no problems to do it in a rhythmical way.

A movement that is visibly non-rhythmical is a symptom of malfunction of the cerebellum. It is important to teach the client to do this exercise in a rhythmical way.

You may help the child to rotate the legs passively by holding just above the ankles and rotating the legs a few times. Then ask the child to do the exercise actively.

In order not to lose the rhythm, this training is best done by repeating the movement very few times, perhaps only three to five times, counting aloud the rhythm of the movement and then pausing and start again. Gradually, the child will learn to repeat the movements more times before losing the rhythm.

This exercise gives a strong stimulation of the cerebellum and activates the cortex. It assists the cerebellum in functioning properly. It helps the client become more aware of the middle line of the body, to control the movement of his feet, and assists the integration of the left and right hemispheres.

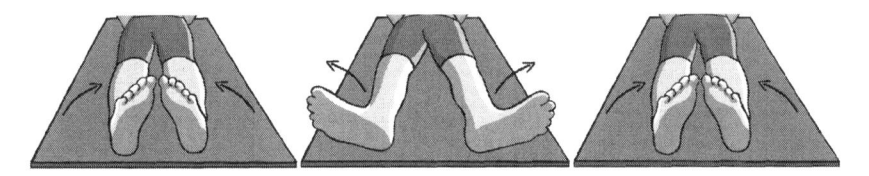

Fig. 35.The windscreen wipers

9. Rolling the bottom from side to side

The client lies on her face with her forehead on her hands and rolls her bottom from side to side. The movement should start from the spine at the level of the lower end of the scapula (T8). If the client has difficulties getting the movement started, as often is the case in children, you can assist by pushing rhythmically from the hip or from the person's waistband.

A. Notice the position of the feet. The big toes should point to each other and the feet should be flat on the mattress. If one or both feet point out to the sides, or if the ankles are so tense that they do not touch the mattress, it indicates a non-integrated Babinski reflex.
B. Notice the position of the shoulders and accessory movements of the head and shoulders. The armpits should be close to the mattress. There should be no accessory movements of shoulders and head. You can subdue the accessory movements by laying a hand on the shoulder blades. Some children have many difficulties in doing this exercise and may make big accessory movements. This usually indicates non-integrated spinal reflexes and poor cooperation between upper and lower body.
C. Notice if the movement is rhythmic or non-rhythmic and if it is symmetric or unsymmetrical. If necessary, correct the movement so it becomes symmetrical. If it is uncoordinated, help by pushing the bottom with your hands so that the body gets the idea of what is needed. You can also instruct the client to roll his bottom three or four times and then rest and repeat. Non-rhythmic movements indicate problems with the cerebellum.
D. When one hip is raised, notice what happens to the leg on the same side. If the leg is bent at the knee, the amphibian reflex has emerged, as it should. If it is stretched, the amphibian reflex has not emerged, which indicates an active spinal galant reflex or asymmetric tonic neck reflex. Correct by instructing the client to bend her knee slightly when raising his hip.

This movement could be made by both small quick deflections in the middle and large deflections to both sides.

By doing this exercise, the client learns to control his back and hips without involving his neck and shoulders. The movement stimulates the cerebellum, especially quick rhythmic movements, and activates the cortex. It gives emotional relaxation.

The movement helps integrate the spinal galant and the spinal Pereze reflexes and establish the amphibian reflex. The rotating movement of the spine stimulates and massages the neuro-lymphatic points on the spine, and the exercise can produce phlegm and dyspnea (difficulty in breathing). Caution is recommended with asthmatic clients.

The exercise also stimulates the spinal pump and improves the circulation of the cerebrospinal fluid.

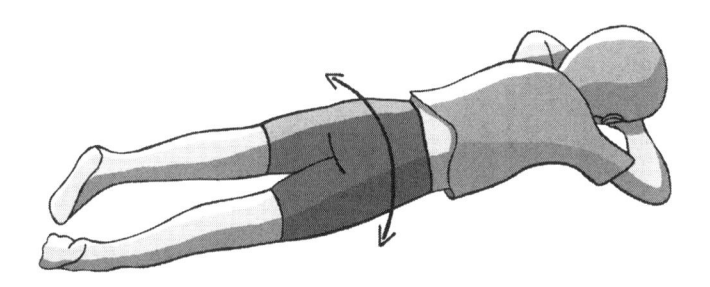

Fig. 36 Rolling the bottom

10. Rocking the body longitudinally in a prone position

The client lies on his face with his hands at the level of his ears and the palms on the mattress, rhythmically pushing from the feet toward the hands. The palms and fingers should be stretched. The head and upper body is raised, while the chin is pulled close to the chest, allowing the neck and head to become an unbroken continuation of the spine. The feet should be at ninety degrees to the legs, and the toes need to be extended.

The client rocks with small rhythmical movements in the longitudinal direction. If this position is difficult to uphold, it may indicate that the Laundau reflex has not emerged.

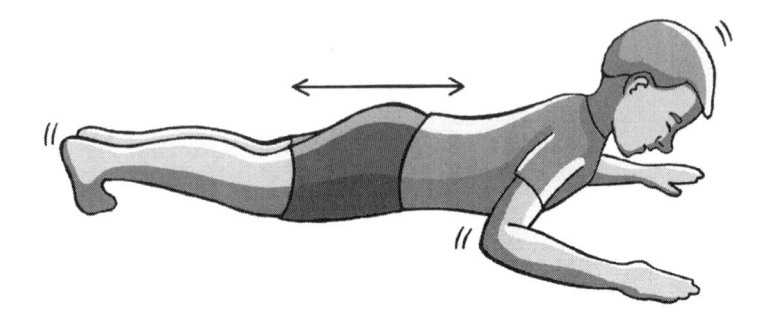

Fig. 37 Longitudinal rocking

Depending on how it is done, the exercise has different effects:

1. Stretching out the feet and letting the palms pull the body rhythmically and exercising the upper arms

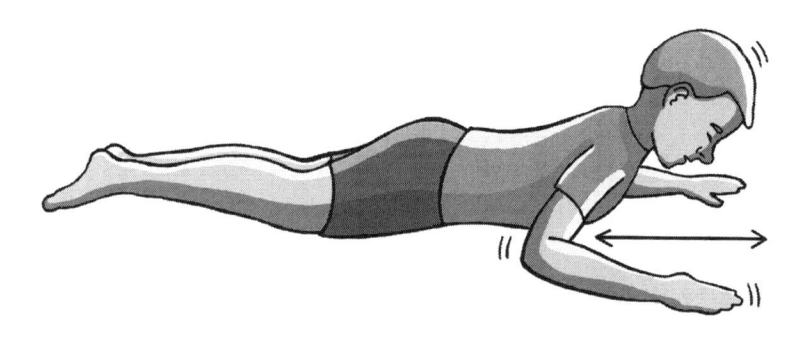

Fig. 38 Longitudinal rocking from the hands

This variation is especially suited for people with weak upper arms and helps them to integrate the symmetric tonic neck reflex, the Babkin reflex, and the grasp and hands pulling reflexes.

2. Pushing backward from the hands toward the toes

The legs should be straight, and the feet need to be at a ninety-degree angle to the legs. This exercise is good for stiff toes and tension in the flexor muscles of the legs. It helps to integrate the tendon guard reflex.

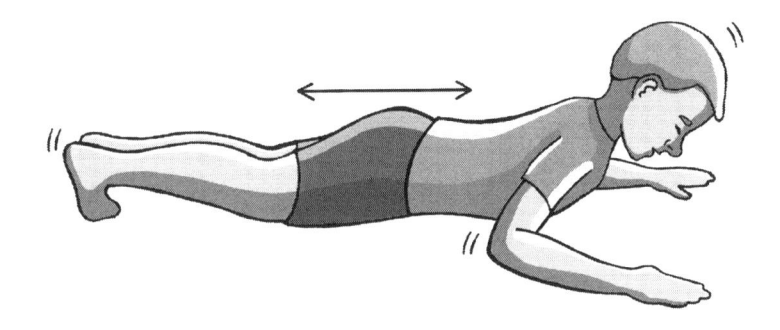

Fig. 39 Longitudinal rocking pushing back

3. In a prone position with the forehead on the hands by rhythmically pushing from the toes in a longitudinal directions toward the hands

Let the shoulders and neck relax and the forehead roll over the hands. This exercise helps to develop coordination between the upper and lower body.

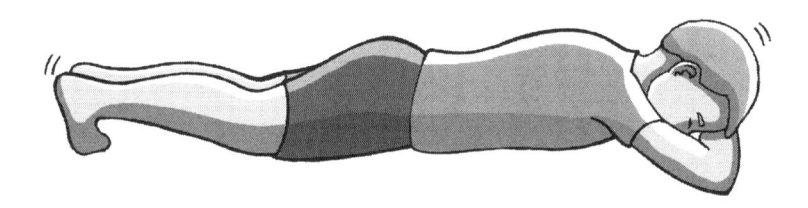

Fig. 40 Longitudinal rocking from the feet

11. Crawling in a prone position

The crawling movement that infants normally make involves both arms and legs in a cross crawl, as has been described above. If the crawling of the child is homolateral (that is, if the leg is bent and the arm is stretched on the same side), this indicates some underlying problem.

Kerstin Linde modified the normal crawling of infants to make it mainly involve the legs and the hips, while the upper body is more passive. The client lies on her face with her forehead on her hands and pulls up one leg to the level of the knee of the other leg. Then she presses down her foot against the mattress and stretches her leg by pushing away. She then repeats this with the other leg. It is important that all toes be involved in this movement. If necessary, put a hard sofa cushion under the stomach to raise the hips and use a soft mattress to crawl on.

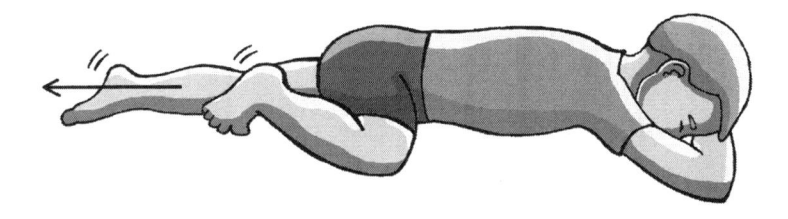

Fig. 41 Crawling on the stomach

An exact crawling movement should have the following characteristics:

A. Both heels should point up to the ceiling and not to the other leg during the whole exercise.
B. All toes of the feet should have contact with the floor when pushing away. If the toes are stiff, this is facilitated if the exercise is done on a soft mattress or if the hips are raised by putting a sofa cushion under the stomach.

C. The bottom must not be raised from the floor, and the hip of the straight leg must be close to it.
D. The toes should be extended and close to the floor and the other leg during the entire exercise.

It is almost impossible for most clients to make an exact crawling movement from the start. Tension in the toes, ankles, and hips must first be loosened.

When the client has learned to crawl exactly on the spot, he can put his hands on his back and crawl forward on the floor. When crawling on a hard surface, the toes should be wet in order to get friction. Mind the risk of sores on the toes in the beginning!

Crawling integrates the crawling reflex, and if it is made in an exact way, it effectively integrates the Babinski reflex. It establishes the cross pattern of the body and helps to integrate the ATNR and the two hemispheres of the brain.

Because this exercise largely involves the hips, it can cause strong emotional reactions and dreams.

Exercises on hands and knees or in a sitting position

Introduction

When working with exercises on hands and knees and in a sitting position, the focus is on the back and the posture. First we look at the posture in an upright position. Is the spine hunched up? Does the client manage to hold up his head or does it drop to the side or forward? The following exercises help the client to control his back, improve posture, eliminate exaggerated or incorrect bending of the back, also improving flexibility of a stiff back.

They are important for the integration of primitive reflexes (e.g., the TLR, the STNR, and the spinal galant reflex). By loosening fixations of the lower back, they may also release repressed emotions.

12. Cat Arches

The client kneels on hands and knees, with straight or slightly bent arms and knees together. Notice the position of the hands, feet, and back. Do the hands point forward, inward, or outward? Are the palms flat on the floor? Are the fingers bent? Are the feet flat to the floor? Is the bending of the back normal or is there an exaggerated lordosis or a slight kyphosis in the lumbar region?

Correct the posture of the client. The hands should be pointed forward and lie flat on the surface, the fingers straight. There should be a slight lordosis of the lumbar region, and the thoracic spine should be straight. If possible, the feet should be flat on the floor.

Ask the client to sag his back slowly and lift his head back and then arch his back up and lower his head. If possible, the arms should be slightly bent at the elbows during the whole exercise.

Fig. 42 Cat arches

Notice if the head follows easily or not. Notice if the lumbar region is rigid and fixed so there is no sagging movement. Notice if the back is too flexible and the lordosis becomes too exaggerated when sagging (like a hammock).

If the hands are directed inward or outward, it indicates weakness of the upper arms and an active symmetric neck reflex (STNR). If the feet are not flat on the floor due to flexion of the knees, it also indicates an active STNR. If the ankles are tense, it indicates a non-integrated Babinski reflex. Hollowed hands and bent fingers indicate an active STNR or active grasp or Babkin reflex.

Stiffness and an inability to sway the back in the lumbar region can indicate active spinal reflexes or an inherent low muscle tone of the back, which has been compensated by rigidity of the spinal muscles. In children, there is often a slight kyphosis in the lumbar back and an inability to sag the lumbar back. This is a compensation for weak upper arms due to an active STNR.

This exercise improves the muscle tone of the back and decreases over flexibility. It helps to integrate spinal reflexes.

13. Somersault rolling

The client kneels on hands and knees in the same position as the preceding exercise. She puts her forehead on the mattress or a soft pillow, her forearms on the mat, and her hands beside her head. To spare the neck, she must put her weight on her forearms and hands, not on her neck. Then she rolls her head from the forehead to the top of the head. This movement is *not* recommended for adults with neck problems but is a wonderful exercise for children.

Do not do this movement in Down syndrome!

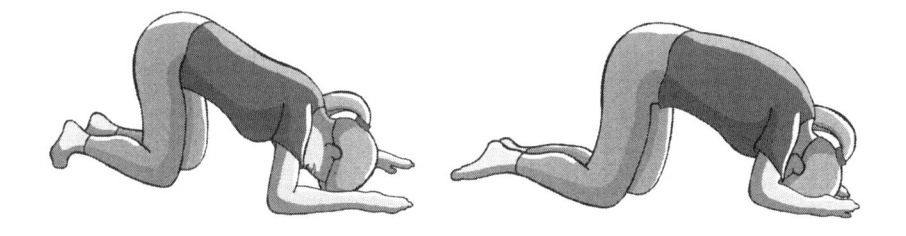

Fig. 43 Somersault rolling

This exercise teaches control of the lumbar region and improves muscle tone of the back. It helps to integrate the tonic labyrinth reflex (TLR) and the symmetric tonic neck reflex (STNR).

14. Rocking on hands and knees

The client sits on his heels and stretches his arms forward. Then he rocks forward until the head is right above his hands, turns, and rocks backward to the heels, bouncing forward again. The arms should be slightly bent, the hands should be directed forward, and the palms and fingers stretched close to the floor. The lumbar region should sag a little, and the thoracic spine should be straight and not arched backward. The ankles should be close to the floor, and the feet must not be raised.

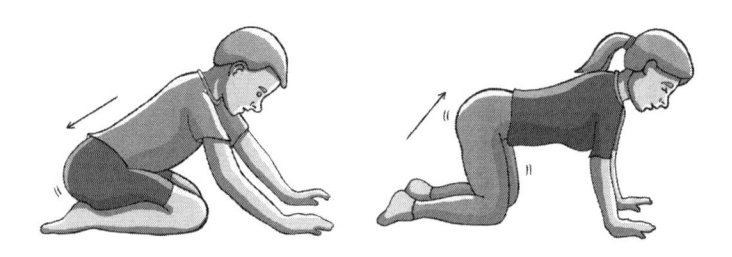

Fig. 44 Rocking on hands and knees

This exercise integrates the STNR. If this reflex is active, there is weakness of the upper arms. This weakness is usually compensated in different ways (e.g., turning the hands inward or outward and locking the elbows, sagging between the shoulder blades, or arching the thoracic or lumbar spine backward). Sometimes the thoracic spine becomes fixed in a slight kyphosis. If the client succeeds in broadening between his shoulder blades and straightening his thoracic spine, the weakness of the upper arms becomes obvious.

15. Sitting on the heels—sagging the lumbar spine forward and arching it backward

The client sits on his heels, slightly leaning forward, sagging and arching his lumbar spine. Only the hips and the lumbar spine should move. To ensure that the shoulders and head are not involved in the movement, the client is asked to raise her arms over her head and keep her head still. The movement should be made *slowly*, with proper deflections forward and backward. If it is made correctly, there should be a cogwheel quality to the movement.

Fig. 45 Sagging the lumbar spine forward and arching it backward

This exercise helps the client to control the lumbar region without involving the head and shoulders, strengthens the muscles of the back, and helps to integrate the spinal reflexes.

GLOSSARY

Accommodation - The ability to change the focus of the eyes from a far to a near distance and back. For clear vision at a near distance, the lens must become spherical, which is brought about by contraction of the ciliary muscle.

Adrenalin – Also called epinephrine, a hormone produced by the adrenal medulla in stress. It increases pulse, blood pressure, and circulation in the skeletal muscles and dilates the pupils.

Amino acids – Building blocks of protein.

Amygdala – A structure in the temporal lobes belonging to the limbic system. It is important for emotional reactions (fear and aggressiveness), memory, and decision making.

Autism Spectrum Disorder (ASD) – Disorders characterized by social deficits and communication difficulties, stereotyped, or repetitive behaviors and interests, and in some cases, cognitive delays. They are described in the American Psychiatric Association's *Diagnostic and Statistical Manual of Mental Disorders, 5th Edition* (DSM-5). They include autism, Asperger's syndrome, and Rett syndrome.

Axon – A nerve fiber that conducts electrical impulses away from the neuron's cell body (the nerve cell body).

Basal ganglia – A part of the brain that, according to Paul MacLean, corresponds to the reptilian brain, it is responsible for our lifelong postural reflexes.

Binocular vision – The ability to direct both eyes to the same object in order to create a three-dimensional picture by fusing the image from each eye into one image.

Brain stem – The region of the brain that connects the cerebrum with the spinal cord, it consists of the midbrain, medulla oblongata, and the pons.

Broca's area – A region in the frontal lobe, usually of the left hemisphere, linked to speech production, language comprehension, grammar, and syntax.

Cerebellum – A region in the back of the brain, between the brain stem and the occipital lobe, that plays an important role in motor control, coordination, attention, and speech.

Cerebral cortex – The outermost layered structure of the brain, it consists of the motor cortex, which initiates voluntary movements, and the sensory cortex, which receives information from the senses: visual cortex, primary auditory cortex, and primary somatosensory cortex. The cerebral cortex plays a key role for consciousness, voluntary movements, perceptual awareness, attention, and language.

Cerebral palsy – A group of permanent nonprogressive movement disorders that cause abnormality of motor function and postural tone. It is caused by damage to the motor control centers of the developing brain and can occur during pregnancy, during childbirth, or after birth, up to about age three.

Ciliary muscle – A ring of smooth muscle that controls accommodation. Parasympathetic activation causes ciliary muscle contraction which decreases the diameter of the ring of the ciliary muscle. The lens becomes more spherical, increasing its power to refract light for near vision. Sympathetic activation of the muscles results in relaxation for vision at a long distance.

Cingulate cortex – Part of the limbic system situated like a border (Latin limbus) around the brain stem and the basal ganglia. According to MacLean, the function of the lobus cingulus is to regulate maternal behavior and play.

Corpus callosum – A bundle of neural fibers that connect the left and right cerebral hemispheres and facilitate communication between the hemispheres.

Cortisol – A hormone produced by the adrenal cortex, it is released in response to stress.

DECT phone – A cordless phone using a single base station to connect one or more handsets to the public telecoms network. The base station functions as a mobile mast and emits radiation when it is plugged in, which makes the DECT phone more harmful than normal cell phones.

Dentate nucleus – A nerve nucleus deep within each cerebellar hemisphere, it is connected by a thick nerve bundle with the frontal lobes and is important for language and control of voluntary movements.

Dopamine – A neurotransmitter of the brain with major functions in the limbic system, prefrontal cortex, and basal ganglia. It is important in reward-motivated behavior and motor control.

Esophoria – When the eyes are turned inward at rest in the absence of visual stimuli for fusion.

Exophoria – When the eyes are turned outward at rest in the absence of visual stimuli for fusion.

Executive functions – Cognitive processes that manage planning, working memory, attention, impulse control, and mental flexibility

Fusion – The ability of the visual cortex to create a three-dimensional picture by fusing the image from each eye into one image.

Fovea – Located in the center of the macula region of the retina, the fovea is responsible for sharp central vision. Also called the central visual field.

GABA – Gamma amino butyric acid, or GABA, is the chief inhibitory neurotransmitter in the mammalian central nervous system. It plays the principal role in reducing neuronal excitability throughout the nervous system.

Kinesthetic sense – The sense of the relative position of different parts of the body and the strength of effort being employed in movement. Same as the proprioceptive sense.

Myelin – An electrically insulating material that forms a layer, the myelin sheath, around the axon of a neuron. The main purpose of a myelin sheath is to increase the speed at which impulses propagate along the myelinated fiber.

Neocortex – The newer six-layered portion of the cerebral cortex, showing the most highly evolved stratification and organization.

Neural pathways – Bundles of axons connecting different areas of the brain.

Neuron – Nerve cell, consisting of a cell body, an axon, and dendrites.

Neurotransmitters – Substance released from the axon terminal, which diffuses across the synapse to transmit the nerve signal to the next neuron.

Occipital lobe – The posterior part of the brain that receives information from the visual sense.

Parietal lobe – The upper central portion of the cerebral hemisphere, between the frontal and occipital lobe, it is concerned with language and general sensory function.

Parkinson's disease – Characterized by symptoms of loss of muscle control, tremor of resting muscles, slowing of voluntary movements, loss of postural reflexes, and poor balance.

Peptide – A molecule consisting of two or more amino acids that combine to make proteins.

Peptidase – Protein-splitting enzyme that breaks peptides into amino acids. Can be found in digestive juice.

Postural reflexes – Lifelong reflexes associated with maintaining normal body posture and mobility. They are necessary for stability and balance in the gravity field, called stability and balance reflexes, and for proper walking and running, called locomotive reflexes.

Prefrontal cortex – The anterior part of the frontal lobes in charge of executive functions.

Primary motor cortex – Located in the posterior part of the frontal lobe, it works in association with other motor areas, the cerebellum, and the basal ganglia to plan and execute movements.

Primitive reflexes – Automatic stereotyped movements controlled from the brain stem. These reflexes are developed during different stages of pregnancy and during the first year of life. They should mature and finally be inhibited by the basal ganglia and integrated into the whole movement pattern of the baby.

Proprioceptive sense – The sense of the relative position of different parts of the body and the strength of effort being employed in movement. Same as the kinesthetic sense.

Purkinje cells – Cells in the cerebellar cortex that use GABA as a transmitter substance.

Reticular activation system (RAS) – A nerve net in the medulla oblongata that receives stimulation from the senses and transmits it to the cerebral cortex. It controls central nervous system activity, including wakefulness, attentiveness, and sleep.

Strabismus – A condition in which the eyes do not point in the same direction, also referred to as a squint. It may cause one or both eyes to turn inward (crossed eyes) or outward (wall eyes).

Suppression – A process in which the brain inhibits the image from one eye in order to avoid double vision in strabismus.

Synapse – A connection between neurons where an impulse is transmitted from one cell to another by a neurotransmitter, which diffuses across the gap to bind with receptors on the postsynaptic cell membrane, causing the nerve signal to continue.

Temporal lobes – Located below the frontal and parietal lobes. Important structures in the temporal lobes are the primary auditory cortex, Wernicke's area, amygdala, and hippocampus. The temporal lobes are involved in auditory processing and comprehending language, emotions, and learning.

Vestibular system – Situated in the vestibulum in the inner ear, it is the sensory system that provides leading contribution about movement and sense of balance.

Visual acuity – Acuteness or clearness of vision.

Wernicke's area – An area in the rear of the left temporal lobe associated with the ability to recognize and understand spoken language.

CV OF HARALD BLOMBERG

I qualified as a doctor in 1972 and worked initially in child psychiatry for several years. In 1976, I started to work in adult psychiatry in order to become a specialist of psychiatry. In 1984, I joined a two-year training program in clinical hypnosis. Among the teachers were several prominent clinical hypnotists from UK and the United States. Peter Blythe, the founder of the Institute for Neuro-Physiological Psychology, was one of the principal teachers of hypnosis. Besides courses in hypnosis, I attended a course in primitive reflexes and learning disabilities.

In 1985, I was introduced to Kerstin Linde, a self-taught body therapist without formal medical education. She was working with rhythmic movements inspired by the movements infants make before they learn to walk. I had been told that she had successfully treated children and grown-ups with severe neurological and other kinds of handicaps. When I met her, I needed to do something about my own motor difficulties, caused by polio in my childhood, and I enlisted as her patient. Her treatment method had a very strong impact on me, and I asked to sit in on her treatments, which she generously allowed. I particularly followed her work with children who suffered from neurological handicaps such as cerebral palsy and saw the most incredible improvements that contradicted all my medical education and experience. I also followed her work with Alzheimer's patients and people with psychosis and other psychological and emotional disturbances. Even in these cases, I was stunned by the positive effects of her treatment. I decided to write a book about her treatment method and started to interview parents of handicapped children who were treated by Kerstin Linde.

In 1982, I had finished my specialist training and started to work as a psychiatric consultant at a psychiatric outpatient clinic. In 1986, I introduced the rhythmic movement training of Kerstin Linde at my clinic, both for neurotic and psychotic patients, with excellent results. We even saw amazing recovery in some cases of protracted schizophrenia. The patients were very grateful and happy for the treatment, but when my superior heard about it, he forbade me to continue this practice. I refused to oblige, and in order to stop me, he had no other alternative but to report me to the National Board of Health and Welfare; an investigation was started in 1988.

In his report to the National Board of Health, my superior argued that he had forbidden me to use this treatment method because "it was not based on reliable experience and scientific evidence and moreover was not generally accepted or especially well known." I wrote a fifty-page report with ten case studies documenting the effects of the treatment, and some twenty of my patients wrote to the National Board of Health to express their appreciation of the treatment. Representatives for the National Board of Health and Welfare also inspected the outpatient clinic where I worked. In its formal report, the National Board of Health and Welfare established that the treatment was "experienced very positively by many patients" and that the "movement treatment was a worthwhile contribution in a situation that had appeared to be deadlocked or stagnant." The board concluded the following: "If every element of the treatment should be called upon for a full scientific documentation, the psychiatric treatment would probably be sterile, which would totally contradict the humanistic values and expressions that psychiatry also had to defend." In the report, I was strongly urged to assist in the initiation of a scientific examination of the treatment method. The board concluded its report by criticizing my superiors for the lack of cooperation between treatment of inpatients and of outpatients and requested a report from the medical superintendent about what measures had been taken to improve this cooperation.

The conflict between my superiors and me lasted more than a year before I was exonerated by the National Board of Health and

Welfare, and by then I had been effectively ostracized by my superiors. My work situation had become impossible, and I decided to resign.

In 1989, I started private practice, and a colleague invited me to introduce the movement training for some severely ill chronic schizophrenic patients, most of them hospitalized for ten years or more at the psychiatric hospital where he worked. I worked with this project two days a week and in my private practice the rest of the time. The next year, in 1991, this work developed into a research program supervised by a professor of psychology at Umeå University. The research program was designed to continue for five years but was unfortunately interrupted in 1994, when I had to quit my work at the psychiatric hospital for private reasons.

However, after two years, in 1993, a report dealing with "short-term changes in chronic schizophrenic patients treated with rhythmic movement therapy" was compiled. The report was an examination paper by two students of psychology. They concluded the following: "The study indicates that the patients treated with movement therapy had displayed the greatest positive changes … Among other things, the changes manifested themselves in the fact that these patients to a greater extent were able to take part in social activities and participate in occupational therapy and their daily tasks in the ward. They had also become more interested in their surroundings."

In 1990, I started to work once every two weeks as a psychiatric consultant at an anthroposophic special school for mentally handicapped youths between the ages of fifteen and twenty-one years. I introduced the rhythmic movement training at this school, where I worked regularly until recently. Some of the students were mentally handicapped; others were diagnosed with autism or ADD. Some of the therapists at the school learned rhythmic movement training, which was offered to those students who were considered to benefit by them. Our experience of this work has been that students with movement disabilities, with learning disabilities due to ADD, and with psychosis are the ones that benefit the most from the rhythmic movement training, while those with autism also needed a gluten- and casein-free diet in order to avoid unnecessary emotional reactions.

In my private practice for adults, I used the rhythmic movement training in combination with movements for inhibition of primitive reflexes. This approach was especially beneficial for patients who as children had suffered from dyslexia and/or ADD. But all patients who agreed to do the rhythmic movements benefited since these movements stimulate the therapeutic process and especially the dreams of the patients.

After quitting my part-time work at the psychiatric hospital in 1994, I started to work full time in my private practice. I got time to start writing a book about the rhythmic movement training that I had planned to write since 1986. My purpose was, among other things, to account for the effects of the rhythmic exercises both in respect of the improvement of motor ability and in respect of stimulation of dreams and psychological development. The book included case reports of treatment with rhythmic training, an attempt to explain its mode of action, and a general discussion of the theoretical basis of scientific medicine. *Helande Liv* was finally published in 1998 by a small publishing house specializing in books about learning disabilities and similar subjects.

Kerstin Linde did not consciously work with primitive reflexes, although she admitted that she observed them. In her experience, the rhythmic exercises were sufficient in order to deal with primitive reflexes. During the 1990s, I attended courses about primitive reflexes taught according to the Peter Blythe and Sally Goddard method. During the first years of 2000, I attended several courses about kinesiology and integration of primitive reflexes, taught by Svetlana Masgutova.

Since 1990, I have taught many lectures and courses in rhythmic movement training for therapists, teachers, and nursing staff. After the publication of my first book, these courses increased in demand. During the following years, I gave courses frequently and regularly. The emphasis of these courses has been on treating children with dyslexia, ADHD, and motor problems.

Based on my experience of teaching courses of rhythmic movement training and what I had learned about primitive reflexes, I wrote the first three comprehensive manuals of rhythmic movement training: (1) rhythmic movement training and primitive reflexes in ADHD; (2) rhythmic movement training and the limbic system; (3) rhythmic movement training in dyslexia. These manuals principally dealt with rhythmic movements as taught by Kerstin Linde but also included information about primitive reflexes and how they could be integrated both by rhythmic exercises and by exercises taught by Svetlana Masgutova.

In 2003 and 2004, I attended camps in Poland, organized by Svetlana Masgutova. I got the opportunity to teach lectures about rhythmic movement training, which aroused great interest and resulted in invitations from Moira Dempsey to teach in Singapore and Malaysia and from Carolyn Nyland to teach in the United States, which I did in 2005.

Since then, I have taught courses frequently and regularly both in Sweden and abroad: the United States, Singapore, Malaysia, Hong Kong, Australia, England, Spain, and Finland. Since 2009, I have mainly taught in Sweden, Spain, Finland, Japan, France, the United States, Germany, Hong Kong, Belgium, and Switzerland.

In 2008, I published another book about rhythmic movements, called *Rörelser Som Helar* (*Movements That Heal*), which was published in English in 2011, with Moira Dempsey. The publishing of that book caused an increasing demand for my courses, and in 2009 I opened a center for rhythmic movement training in Stockholm, where I teach the rhythmic movement courses and treat patients, mainly children, who suffer from problems with attention, mobility, reading and writing, or who have been diagnosed with autism or Asperger's.

In 2010, I published the book *Autism: En Sjukdom Som Kan Läka* (*Autism: A Disease That Can Heal*), which deals with the environmental causes of autism and treatment of autism with diet, food supplements, and rhythmic movement training, among many other things.

In 2011, I published two booklets, one about central stimulants and one about dyslexia.

In 2014, I published a booklet titled "Gluten-Related Disorder in Children and Adults."

I presently offer the following courses:

Rhythmic Movement Training and Primitive Reflexes in ADHD/ADD (two days)
Rhythmic Movement Training, Emotions and Inner Leadership (two days)
Rhythmic Movement Training in Reading and Writing Difficulties (two days)
Rhythmic Movement Training in Preschool (two days)
Rhythmic Movement Training for Physiotherapists and Massage Therapists (two days)
Rhythmic Movement Training, Dreams and Inner Healing (two days)
Rhythmic Movement Training and Diet in Autism (two days)
Rhythmic Movement Training in cerebral palsy (two days)

REFERENCES

1. Andrew Bridges, Associated Press, 060104
2. Peter Breggin: *Talking back to Ritalin*, Da Capo Press, 2001, page 259
3. Peter Breggin, page 68
4. Peter Breggin, page 24
5. BBC, Panorama, *What next for Craig*, 12 november 2007,
6. Ibid.
7. MTA Cooperative Group (2007), Secondary Evaluations of MTA 36-Month Outcomes: Propensity Score and Growth Mixture Model Analyses, *Journal of the American Academy of Child & Adolescent Psychiatry*, Volume 46(8), August 2007
8. Aarskog, D. Fevang, F., Klöve, H., Stöa, K.,och Thorsen, T., (1977) The effect of stimulant drugs, dextroamphetamine and metylphenidate, on secretion of growth hormone in hyperactive children. *Journal of Pediatrics, 90,* 136-139
9. Nasrallah, H., Loney, J., Olson, S., Mc-Calley-Whitters, M., Kramer, J., and Jacoby, C. (1986) Cortical atrophy in young adults with a history of hyperactivity in childhood. *Psychiatry Research* 17;241-246
10. Swanson, J.M., Cantwell, D., Lerner M., Mc Burnett, K., Pfiffner, L., and Kotkin, R. (1992)Treatment of ADHD: Beyond Medication. *Beyond Behavior 4:No 1* sid. 13-16 och 18-22.
11. Drug Enforcement Administration (DEA). (1995 b, October 20) Metylphenidate; DEA press release
12. Lambert, N., & Hartsough, C.S.(1998). Prospective study of tobacco smoking and substance dependence among samples of ADHD and non-ADHD subjects. *Journal of Learning Disabilities 31,* 533-534
13. Breggin, page 71 – 72
14. Melega, W.P., Raleigh, M.J., Stout, D., B., Lacan, G., Huang, S., C., & Phelps, M. E. (1997b) Recovery of Striatal dopamine function after acute amphetamine- and methamphetamine-induced neurotoxicity in the vervet monkey. *Brain Research, 766,* 113-20.

[15] Raine ADHD Study report: Long-term outcomes associated with stimulant medication in the treatment of ADHD in children. Government of Western Australia, Department of Health (http://www.health.wa.gov.au/publications/documents/MICADHD_Raine_ADHD_Study_report_022010.pdf
[16] Kort om ADHD hos barn och vuxna. En sammanfattning av Socialstyrelsens kunskapsöversikt, 2004.
[17] Robert Winston, *The Human Mind*, International Edition, November 23, 2004, page 78
[18] Paul D. MacLean: *The Triune Brain in Evolution*, Plenum Press 1990
[19] Jean Ayres, *Sensory Integration and the Child*, WPS, 2000, page 72
[20] Breggin, page 217
[21] Breggin, page 86
[22] Breggin, page 34-35
[23] Breggin, page 40
[24] Jaffe, J.H. (1995) Amphetamine (or amphetaminelike)-related disorders. In H.I. Kaplan and Saddock, B. (Eds) *Comprehensive textbook of psychiatry, IV*, pp. 791-799. Baltimore: Williams & Wilkins.
[25] Associated Press, Jan. 4, 2006
[26] Eli Lilly, "Poster" presented at a conference in Florence, August 28 2007
[27] The Medicines and Healthcare Products Regulatory Agency MHRA, *Preliminary Assessment Report*, December 2005
[28] FDA, Report, March 2006,
[29] Swedish daily Metro, January 23 2006
[30] Swedish daily Svenska Dagbladet, June 12 2007
[31] According to an article by Bertil Wosk, Pelle Randberg: Konstgjort sötningsmedel ett hot mot vår hälsa, *Näringsråd och Näringsrön* 2001, 6.
[32] New fears over additives in children´s food by Felicity Lawrence, *The Guardian* May 8 2007
[33] Mona Nilsson, Maria Lindblad; *Spelet om 3G*, Medikament Faktapocket 2005
[34] Salford LG m.fl: Nerve Cell Damage in Mammalian Brain after Exposure to Microwaves from GSM Mobile Phones, *Environmental Health Perspectives* 2003:11
[35] Divan el al.: Prenatal and Postnatal Exposure to Cell Phone Use and Behavioral Problems in Children; *Epidemiology* Vol 19, Number 4, July 2008
[36] Jan Wållinder, *Transmittorn* nr 7
[37] Kort om ADHD hos barn och vuxna.
[38] Paul D. MacLean, page 23

[39] Sally Goddard, *Reflexes, Learning and Behavior*, Fern Ridge Press, 2002, page 89
[40] *En Bok om Hjärnan*, Tiden, Rabén Prisma, 1995. Page 149
[41] Svetlana Masgutova with Nelly Akhmatova: *Integration of Dynamic and Postural Reflexes into the Whole Body Movement System*, Warsaw 2004
[42] Frank A. Middleton and Peter L. Strick: Cerebellar Projections to the Prefrontal Cortex of the Primate, *Journal of Neuroscience* 21(2):700-712
[43] Torleiv Höien, Ingvar Lundberg, *Dyslexi*, Natur och Kultur 1999, sid. 182
[44] See James Purdon Martin: *The Basal Ganglia and Posture*, London, Pitman Medical Publishing Co. Ltd, 1967
[45] Ibid.
[46] Bryan Jepson: *Changing the Course of Autism*, page115, Sentient Publications, 2007
[47] Svetlana Masgutova
[48] Sally Goddard, page 142
[49] Jean Ayres, *Sensory Integration and the Child*, WPS, 2000, page 39
[50] Ibid., page 55
[51] Paul MacLean page 327
[52] Berit Heir Bunkan, *Muskelspänningar*, Universitetsforlaget Oslo 1980, page 43
[53] Jean Ayres, page 137
[54] Paul MacLean, page 396
[55] Elkhonin Goldberg: *The Executive Brain*, Oxford University Press 2001, page 34
[56] Ibid., page 36
[57] Frank A. Middleton och Peter L.Strick: Basal-ganglia Projections to the Prefrontal Cortex of the Primate, *Cerebral Cortex*, Vol. 12, Nr 9
[58] Elkhonin Goldberg, page 139
[59] Bryan Jepson, page 35
[60] Ibid., page 25
[61] Functional impact of global rare copy number variation in autism spectrum disorders. *Nature*, Published online 09 June 2010
[62] Geier DA, et al. (2009). A prospective study of prenatal mercury exposure from maternal dental amalgams and autism severity *Acta Neurobiologiae Experimentalis*, 69:189-197.
[63] Robert Kennedy Jr, Autism, Mercury and Politics, *Boston Globe* July 1 2005
[64] William Shaw: *Biological Treatments for Autism and PDD*, 2002, page 5 and 102
[65] Wakefield AJ. et al. (1998). Ileal-lymphoid-nodular hyperplasia, non-specific colitis, and pervasive development disorder in children. *Lancet* 1998 Feb 28;351(9103):637-41.

[66] Bryan Jepson, 2007, pages 82-86.
[67] Reichelt. K.L.et al. Childhood Autism: A Complex Disorder. *Biol. Psychiatry* 21, 1986
[68] Sapone A. et al. Spectrum of gluten-relate disorder: consensus on a new nomenclature and classification. *BMC Medicine* 2012 10:13
[69] Andrew Wakefield, *Waging war on the autistic child*, Skyhorse Publishing 2012, page 53
[70] Knivsberg A M, Reichelt KL, Höjen T et al. : A randomised, controlled study of dietary intervention in autistic syndromes. *Nutritional Neuroscience* 2002: 13:87-100
[71] Whiteley P, Haracopos D, Knivsberg AM et al. "The ScanBrit randomized, controlled, single-blind study of a gluten-and casein-free dietary intervention for children with autism spectrum disorders." *Nutritional neuroscience* 2010; 13;87-100
[72] www.bioinitiative.org
[73] Bryan Jepson, pages 172 and 252
[74] Amy Yasko, *Autism: Pathways to Recovery*, 2009 Neurological Research Institute, page 68
[75] Ibid., page 59
[76] David H Ingvar, Abnormal Distribution of Cerebral Activity in Chronic Schizophrenia. *Perspectives in Schizophrenia Research*, New York 1980
[77] Reichelt, K.L.et al., Urinary Peptides in Schizophrenia and Depression. *Stress Medicine 1*, 1985
[78] Dohan F.C. Grasberger J.C.: Relapsed Schizophrenics: Earlier Discharge from Hospital after Cereal-Free, Milk-Free Diet, *American Journal of Psychiatry*, 130
[79] Mårten Kalling, *Gap junctions, Kanalerna mellan celler*, Unpublished manuscript June 2007
[80] Robert G. Heath: Modulations of Emotions with a Brain Pacemaker, *The Journal of Nervous and Mental Disease*, No 5 1977
[81] Robert G. Heath: Gross Pathology of the Cerebellum in Patients Diagnosed and Treated as Functional Psychiatric Disorders, *The Journal of Nervous and Mental disease* No 10 1979
[82] Paul MacLean: page 527-534
[83] According to an article byAnna-Lena Haverdahl: Var sjätte dog efter lobotomi. Svenska Dagbladet April 30 2007
[84] Paul MacLean: page 529
[85] Mats Lindqvist & Gerd Pettersson: Rytmiska rörelseterapi med kroniskt schizofrena patienter, Examensarbete 20 poäng, Umeå Universitet, 1993
[86] Rodney P Ford, The Gluten Syndrome: A Neurological Disease, *Medical Hypotheses* 73, No 3 (September 2009)

[87] Myrberg, Mats editor: *Att skapa consensus om skolans insatser för att motverka läs- och skrivsvårigheter*
[88] Boder, E. 1973. Developmental dyslexia: A diagnostic approach based on three atypical reading-spelling patterns. *Developmental Medicine and Child Psychology, 15,* 663-687
[89] Gjessing, H. 1977. *Dysleksi.* Oslo: Universitetsförlaget
[90] Aaron, P.G. 1978. *Dyslexia, an imbalance in cerebral information processing strategies.* Perceptual and Motor Skills, 47, 699-706
[91] Hellige, Joseph B: *Hemispheric Asymmetri*, page 36. Harvard University Press, 1993
[92] According to Kandel, Schwartz & Jessel: *Principles of Neural Science,* 3d edition 1991,
[93] Elkhonin Goldberg, page 49
[94] Lasse Müller: *Optometri vid läs-och skrivsvårigheter,* 2006.
[95] David Cook: *When your child struggles.* Invision Press, Atlanta, 2004
[96] Ibid.
[97] Svetlana Masgutova, page 66
[98] M McPhillips, P G Hepper, G Mulhem: Effects of replicating primary-reflex movements on specific reading difficulties in children: a randomised, double-blind, controlled trial. *The Lancet* Vol. 355, No 9203, page 537-541
[99] Mc Philips M, Jordan-Black J.-A., Primary reflex persistence in Children with reading difficulties (dyslexia): A cross-sectional study, *Neuropsychologia* 2006
[100] Gesell, Arnold et al.. *Vision: Its Development in Infant and Child,* 1998
[101] Johansen, Kjeld V: *Lyd, hoerelse og sprogudvikling,* 1993,
[102] Moats, L. (1996) Phonological spelling errors in the writing of dyslectic children. *Reading and Writing: An Interdisciplinary Journal, 8,* 105-119
[103] Hamilton, Gregory and others : Psychiatric Symptoms and Cerebellar Pathology. *Am J Psychiatry* 14:10, October 1983
[104] Steen Larsen: *Laesningens mysterium,* Hellerup 1996 page 112
[105] Elkhonin Goldberg, page 24
[106] Ibid., page 114
[107] Berg, Lars-Eric och Cramér, Anna: *Hjärnvägen till inlärning,* Natur och Kultur 2011
[108] McPhilips M, Jordan-Black J.-A

Printed in Great Britain
by Amazon